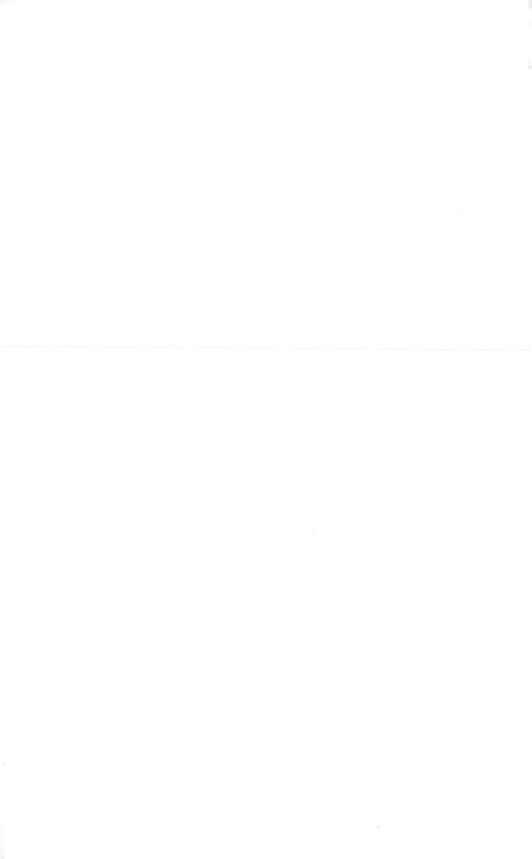

RECURSIVENESS

RECURSIVENESS

Samuel Eilenberg

COLUMBIA UNIVERSITY
NEW YORK

Calvin C. Elgot

IBM, INC.
YORKTOWN HEIGHTS, NEW YORK

ACADEMIC PRESS   New York and London   1970

ACADEMIC PRESS, INC.
111 Fifth Avenue, New York, New York 10003

United Kingdom Edition published by
ACADEMIC PRESS, INC. (LONDON) LTD.
Berkeley Square House, London W1X 6BA

LIBRARY OF CONGRESS CATALOG CARD NUMBER: 70-117099
AMS 1968 Subject Classification No. 0270

PRINTED IN THE UNITED STATES OF AMERICA

Contents

PREFACE   vii

CHAPTER I

Preliminaries

1. Functions   1
2. Relations and Partial Functions   2
3. Categories   3
4. Admissible Subcategories of $\mathscr{R}(X^{*})$ 5
5. Free Monoids   8

CHAPTER II

Primitive Recursive Functions

1. Exponentiation   11
2. Primitive Categories   13
3. Induction   15
4. The Case $W = N$ 17
5. Dominated Minimization   18
6. Primitive Recursive Sets   20
7. Primitive Recursive Isomorphisms   23
8. Various Induction Schemes   24

CHAPTER III

Recursive Isomorphisms

1. Definitions and Results   29
2. Description of the Bijection $a : W \to N$ 32
3. Proof of Condition (i)   33
4. Translated Results   37

v

CHAPTER IV

Recursive Relations

1. Distinguished Subcategories of $\mathscr{R}(W^{*})$ 39
2. Further Consequences of the Axioms   42
3. The Main Theorem   43
4. Proof of the Main Theorem   45

CHAPTER V

Recursive Functions

1. Repetition   51
2. Minimization   53
3. An Inversion Theorem   56
4. Distinguished Subcategories of $\mathscr{P}$ 58

CHAPTER VI

Recursively Enumerable Sets

1. $\mathscr{A}$-Enumerable Sets   63
2. Γ-Enumerable Sets   65
3. Axioms for Enumerability   67
4. Conjugation   70

APPENDIX A

A Recursive Function
Which Is Not Primitive Recursive

73

APPENDIX B

Degrees

83

BIBLIOGRAPHY   85

INDEX   87

INDEX OF NOTATION   89

Preface

The notion "recursive function" or its equivalent "computable function" was formulated in the mid-1930's. Its theory has since been undergoing vigorous development. By now, there are a number of excellent sources of information on the subject (Bibliography, p. 85). The reader will find in these books not only a comprehensive development of the mathematical theory but also an account of the intuitive and historical background of the subject.

More recently, there has been much interest in reexamining the fundamentals of the subject. On the one hand, it appears that a more algebraic, less arithmetic point of view has been sought and, on the other hand, a theory of programs for digital computers appears to be the long-range aim. Interestingly enough, these two motivations appear to be compatible with each other.

This short monograph is a contribution to this latter activity. Its algebraic flavor is more or less obvious. Its connection with computer science is less obvious. The reader who has some familiarity with finite automata theory or mathematical linguistics will note that obvious counterparts of some of the operations which play a central role in our discussion also play central roles in those studies.

In content, the monograph is quite elementary. For example, "Gödelization" is not introduced into our discussion at all. As a consequence, certain universality theorems are not included.

Because of the novelty of the approach, known theorems often appear in unfamiliar guises. Because of this we have not attempted to assign authorship to the theorems.

A cross reference of the type "Proposition IV,6.1" refers to Proposition 6.1 of Chapter IV. The IV is omitted if the reference is in Chapter IV. The symbol ∎ is used in place of a period at the end of a proof, or at the end of theorems or corollaries not followed by a proof.

RECURSIVENESS

CHAPTER **1**_____

Preliminaries

1. Functions

Given sets X and Y we shall use the notation

$$f : X \longrightarrow Y \quad \text{or} \quad X \xrightarrow{f} Y$$

to denote a function defined for all $x \in X$ and with values in Y.
If $g : Y \to Z$ is another function, we denote by $fg : X \to Z$ the
composition

$$fg : X \xrightarrow{f} Y \xrightarrow{g} Z$$

Thus for each $x \in X$ we have

$$(fg)(x) = g(f(x))$$

With this notation, the parentheses around x may be omitted,
but the other pair must be kept. An alternate notation would write
the function symbol after the variable. In this notation, xfg may
be written unambiguously without parentheses. The last notation
has the following added advantage. The element $x \in X$ may be
interpreted as a function $x : I \to X$ where I is some set containing
a single element and the value of the function is x. Then xfg is
the composition of functions

$$I \xrightarrow{x} X \xrightarrow{f} Y \xrightarrow{g} Z$$

In the sequel, we shall use either of the two notations according to convenience. In complicated formulas we shall use the first notation (putting the function symbol to the left of the variable) and insert the necessary parentheses.

2. Relations and Partial Functions

For any set X we denote by $\hat{X}$ the power set of X, that is, the set of all subsets of X (including the empty subset $\varnothing$). Each element x of X will be regarded as a subset consisting of a single element. With this convention we have $X \subset \hat{X}$.

A *relation* f from X to Y, written $f : X \to Y$, is a function

(2.1) $f : \hat{X} \to \hat{Y}$

which is completely additive, that is, which satisfies

$$f\left(\bigcup A_i\right) = \bigcup f(A_i)$$

where $\{A_i \mid i \in I\}$ is a family of subsets of X indexed by some set I and $\bigcup$ stands for the union. Note that $f(\varnothing) = \varnothing$. In view of this additivity, f is completely determined by its values on single element subsets so that f may be viewed as a function

$$f : X \to \hat{Y}$$

If for each $x \in X$ the set $f(x)$ has a single element, then f is a function. Thus, relations are generalizations of functions and as much as possible, we shall use the same notation. Composition of relations is defined in the obvious way using the original definition (2.1).

The *graph* γf of a relation $f : X \to Y$ is defined as a subset of $X \times Y$ as follows:

$$\gamma f = \{(x, y) \mid y \in fx\}$$

Clearly, every subset of $X \times Y$ is the graph of some relation from X to Y; thus, γ is a bijection between the relations $f : X \to Y$ and the subsets of $X \times Y$. This leads many authors to define a relation

$X \rightarrow Y$ as a subset of $X \times Y$. We prefer to separate the notion of a relation from that of its graph for the following reason. A subset of the product $X \times Y \times Z$ (treated associatively) will be the graph of a relation $X \rightarrow Y \times Z$ and also of a relation $X \times Y \rightarrow Z$. These two relations may have quite different properties.

A relation $f : X \rightarrow Y$ is called a *partial function* if for each $x \in X$ the set fx has at most one point.

The *domain* of a relation $f : X \rightarrow Y$ is defined as

$$\mathrm{Dom}\, f = \{x \mid x \in X, \quad fx \neq \varnothing\}$$

EXERCISE 2.1. *Given relations* $f : X \rightarrow Y$, $g : Y \rightarrow Z$, *verify*

$$\gamma(fg) = \{(x, z) \mid (x, y) \in \gamma f, \quad (y, z) \in \gamma g \quad \text{for some} \quad y \in Y\}$$

3. Categories

A *category* $\mathscr{A}$ consists of the following data.

(3.1) *Objects.* A class $\mathrm{Obj}\, \mathscr{A}$, the elements of which are called *objects* of $\mathscr{A}$.

(3.2) *Morphisms.* A funtion which to every ordered pair (A, B) of objects of $\mathscr{A}$ assigns a set $\mathscr{A}(A, B)$; the elements $f \in \mathscr{A}(A, B)$ are called *morphisms* from A to B. We frequently write $f : A \rightarrow B$ or $A \xrightarrow{f} B$ instead of $f \in \mathscr{A}(A, B)$.

(3.3) *Composition.* An operation of *composition* which to each consecutive pair of morphisms

$$A \xrightarrow{f} B \xrightarrow{g} C$$

assigns a morphism

$$fg : A \rightarrow C$$

These primitive terms are subject to the following axioms.

(3.4) *Associativity.*   Given morphisms

$$A \xrightarrow{f} B \xrightarrow{g} C \xrightarrow{h} D$$

we have

$$(fg)\,h = f(gh)$$

(3.5) *Identity.*   For each object A, there exists a morphism

$$1_A : A \to A$$

such that $1_A f = f$ for every morphism $f : A \to B$ and $g 1_A = g$ for every morphism $g : C \to A$.

We note that 1_A is unique. Indeed, if $1_A' : A \to A$ were another such identity, then $1_A = 1_A 1_A' = 1_A'$.

A *subcategory* $\mathscr{A}'$ of a category $\mathscr{A}$ is a category such that

(3.6)   Obj $\mathscr{A}'$ is a subclass of Obj $\mathscr{A}$.

(3.7)   For A, B in $\mathscr{A}'$, the set $\mathscr{A}'(A, B)$ is a subset of $\mathscr{A}(A, B)$.

(3.8)   The composition of morphisms in $\mathscr{A}'$ is that of $\mathscr{A}$.

(3.9)   For A in $\mathscr{A}'$, 1_A is in $\mathscr{A}'(A, A)$.

A subcategory $\mathscr{A}'$ of $\mathscr{A}$ is said to be a *full* subcategory if $\mathscr{A}'(A, B) = \mathscr{A}(A, B)$ for A, B in $\mathscr{A}'$. Thus, a full subcategory is obtained by restricting the class of objects.

Most frequently we shall write A for the identity morphism $1_A : A \to A$.

An example of a category is the category $\mathscr{R}$ of relations. The objects are sets and the morphisms in $\mathscr{R}(X, Y)$ are all the relations $f : X \to Y$, with composition being composition of relations. Partial functions yield a subcategory $\mathscr{P}$ of $\mathscr{R}$, and functions yield a subcategory $\mathscr{F}$ of $\mathscr{P}$.

All these are examples of *large* categories in which the objects form a class which is not a set (in the sense of Bernays–Godel–von Neumann set theory). Categories in which the objects form a set are called *small*. In the sequel we shall encounter only small categories.

4. Admissible Subcategories of $\mathcal{R}(X^\cdot)$

Given a set X we shall denote by $X^\cdot$ the sequence of sets $\{X^r\}$, $r \geqslant 0$, where X^0 is the set consisting of a single point (say the number 1), and where $X^1 = X$ and $X^{i+1} = X^i \times X$. We agree to identify $X^p \times X^q$ with X^{p+q} for $p \geqslant 0$, $q \geqslant 0$.

The category $\mathcal{R}(X^\cdot)$ is now defined as the full subcategory of $\mathcal{R}$ determined by the objects X^r, $r \geqslant 0$. Thus the objects of $\mathcal{R}(X^\cdot)$ are the sets X^r and the morphisms are relations $f : X^r \to X^s$, $r \geqslant 0$, $s \geqslant 0$. The categories $\mathcal{P}(X^\cdot)$ and $\mathcal{F}(X^\cdot)$ are defined as

$$\mathcal{P}(X^\cdot) = \mathcal{R}(X^\cdot) \cap \mathcal{P}, \qquad \mathcal{F}(X^\cdot) = \mathcal{R}(X^\cdot) \cap \mathcal{F}$$

and are full subcategories of $\mathcal{P}$ and $\mathcal{F}$, respectively.

In dealing with the category $\mathcal{R}(X^\cdot)$ and its various subcategories, the following notational convention will be useful. In writing $(a, b, c) \in X^{p+q+r}$, it will be automatically understood that $a \in X^p$, $b \in X^q$, and $c \in X^r$. Thus, the particular form in which the exponent of X is written has significance by agreement.

Another notational convention that is useful is the following. Given subsets $A \subset X^p$, $B \subset X^q$, we shall denote by (A, B) the subset $A \times B$ of X^{p+q}. This generalizes the notation (x, y) when $A = x$, $B = y$ are single element sets.

A subcategory $\mathcal{A}$ of $\mathcal{R}(X^\cdot)$ is called *admissible* if it satisfies the following axioms.

(4.1)   The objects of $\mathcal{A}$ are all the objects X^r, $r \geqslant 0$ of $\mathcal{R}(X^\cdot)$.

(4.2)   *Cylindrification.* If $f : X^r \to X^s$ is a relation in $\mathcal{A}$, then so is the relation $X \times f : X^{1+r} \to X^{1+s}$, defined by

$$(X \times f)(x, y) = (x, fy)$$

for $x \in X$, $y \in X^s$.

(4.3)   *Transposition.*   The function $\Theta_k : X^k \to X^k \, (k \geqslant 2)$, defined by

$$\Theta_k(x, y, z) = (y, x, z)$$

for $x, y \in X$, $z \in X^{k-2}$, is in $\mathscr{A}$.

(4.4)   *Diagonal.*   The function $\varDelta : X \to X^2$, defined by $\varDelta x = (x, x)$, is in $\mathscr{A}$.

(4.5)   *Projection.*   The unique function $\Pi : X \to X^0$ is in $\mathscr{A}$.

Clearly $\mathscr{P}(X^{\bullet})$ and $\mathscr{F}(X^{\bullet})$ are admissible subcategories of $\mathscr{R}(X^{\bullet})$. For each integer $n \geqslant 0$, let $[n]$ denote the set $\{1, ..., n\}$. Thus in particular, $[0] = \varnothing$. Given a function $f : [p] \to [q]$, we define the *logical function* $f^{\#} : X^q \to X^p$ induced by f by setting

$$f^{\#}(x_1, ..., x_q) = (x_{f1}, ..., x_{fp})$$

Clearly the functions Θ_k, $\varDelta$, and Π are logical functions; indeed, Θ_k is induced by the permutation $[k] \to [k]$ that interchanges 1 and 2 and holds fixed all other elements while $\varDelta$ and Π are induced by the unique functions $[2] \to [1]$ and $\varnothing \to [1]$. The functions Θ_k, $\varDelta$, and Π will be called the *basic logical functions.*
In the rest of this section, it will be assumed that $\mathscr{A}$ is an admissible subcategory of $\mathscr{R}(X^{\bullet})$.

PROPOSITION 4.1.   *All logical functions are in $\mathscr{A}$.*

Proof.   Let $f^{\#} : X^q \to X^p$ be induced by $f : [p] \to [q]$. From (4.2) and (4.3) we deduce the conclusion in the case when $p = q$ and f is a transposition of two consecutive integers. Since $(fg)^{\#} = g^{\#}f^{\#}$, it follows that the conclusion holds when $p = q$ and f is a permutation.

An arbitrary function $f : [p] \to [q]$ is the composition of a surjection followed by an injection, and it suffices to prove the conclusion in each of these cases.

A surjection $f : [p] \to [q]$ with $p > q$ is the composition of surjections of the type $f : [r + 1] \to [r]$. By an appropriate permutation, such a surjection reduces to the function satisfying $fi = i$

for $i \leqslant r$ and $f(r+1) = r$. In this case, $f^{\#} = X^{r-1} \times \varDelta$. Thus, $f^{\#}$ is in $\mathscr{A}$ in this case.

An injection $f : [p] \rightarrowtail [q]$ is the composition of injections of the type $f : [r] \rightarrowtail [r+1]$ and by an appropriate permutation is reduced to the inclusion mapping $f : [r] \rightarrow [r+1]$. In this case $f^{\#} = X^r \times \Pi$ ∎

PROPOSITION 4.2. *If $f : X^r \rightarrow X^s$ and $g : X^t \rightarrow X^u$ are in $\mathscr{A}$, then so is*

$$f \times g : X^{r+t} \rightarrow X^{s+u}$$

defined by

$$(f \times g)(x, y) = (fx, gy)$$

for $x \in X^r$, $y \in X^t$.

Recall that for convenience we write (fx, gy) instead of $(fx) \times (gy)$.

Proof. If $r = s$ and f is the identity, then $f \times g = X^r \times g$ and the result follows by r-fold cylindrification. If $s = u$ and g is the identity, then $f \times g = f \times X^s$ and the result follows by permutation of coordinates (Proposition 4.1) and s-fold cylindrification. In the general case, $f \times g$ is the composition

$$X^{r+t} \xrightarrow{f \times X^t} X^{s+t} \xrightarrow{X^s \times g} X^{s+u} \quad ∎$$

Note that if f and g are partial functions, then so is $f \times g$ and $\mathrm{Dom}(f \times g) = \mathrm{Dom}\, f \times \mathrm{Dom}\, g$. Thus if f and g are functions; so is $f \times g$.

PROPOSITION 4.3. *If $f : X^r \rightarrow X^s$ and $g : X^r \rightarrow X^t$ are in $\mathscr{A}$, then so is*

$$\langle f, g \rangle : X^r \rightarrow X^{s+t}$$

defined by

$$\langle f, g \rangle x = (fx, gx)$$

Proof. $\langle f, g \rangle$ is the composition

$$X^r \xrightarrow{\Delta_r} X^{r+r} \xrightarrow{f \times g} X^{s+t} \quad \blacksquare$$

Note that $\mathrm{Dom}\langle f, g \rangle = \mathrm{Dom}\, f \cap \mathrm{Dom}\, g$ and if f and g are partial functions, then so is $\langle f, g \rangle$. Thus if f and g are functions, then so is $\langle f, g \rangle$.

The operations described in Propositions 4.2 and 4.3 are associative and thus lead to operations on several factors.

5. Free Monoids

In Section 4 we considered admissible subcategories of the category $\mathscr{R}(X^{\cdot})$. In practice, the set X will not be arbitrary, but will have an algebraic structure. In fact, in much of the sequel, X will be a free monoid on a finite base.

To describe such a free monoid, we consider a finite set Σ with distinct elements $\{\sigma_1, ..., \sigma_n\}$, $n > 0$. The elements σ_i are called *letters* and Σ is called an *alphabet*. A sequence w of k letters is called a *word* of length k. If u is a word of length l, then the word wu of length $k + l$ is formed by concatenation. This multiplication is associative and thus every word w of length k is the product of k words of length 1; that is, a product of k letters. There is one word of length 0. This word is denoted by 1 and is a unit element for the multiplication. The monoid thus obtained is denoted by Σ^*. We shall also use the letter W to denote this monoid.

In the sequel, admissible subcategories of $\mathscr{R}(W^{\cdot})$ will be studied. Certain functions will occur repeatedly. They are

Left successors $L_i : W \to W$, $L_i w = \sigma_i w$, $1 \leqslant i \leqslant n$

Right successors $R_i : W \to W$, $R_i w = w\sigma_i$, $1 \leqslant i \leqslant n$

Unit $U : W^0 \to W$, $U1 = 1$

Reversal $\rho : W \to W$, $\rho 1 = 1$, $\rho\sigma = \sigma$, $\rho(wu) = (\rho u)(\rho w)$

The special case $n = 1$ will be of interest. The monoid W consists then of the powers σ^k, $k \geqslant 0$ of the single letter σ. Shifting to additive notation, we may identify W with the set N of all nonnegative integers. The distinction between left and right successors disappears and the single successor function is

$$S : N \rightarrow N$$

$$Sx = x + 1$$

The length of a word $w \in W$ will usually be denoted by $|w|$. If $W = N$, then $|n| = n$.

CHAPTER II

Primitive Recursive Functions

1. Exponentiation

A function

$$l : W^{1+r} \to W^r$$

is called a *left action* if it satisfies

(1.1) $$l(1, x) = x$$

(1.2) $$l(uv, x) = l(u, l(v, x))$$

for all $u, v \in W$ and all $x \in W^r$.

The formulas above may be reinterpreted as follows. For each $w \in W$, define the function

$$l_w : W^r \to W^r$$
$$l_w(x) = l(w, x)$$

Then (1.1) and (1.2) become

(1.1') $$l_1 = W^r$$

(1.2') $$l_{uv} = l_v l_u$$

Thus l viewed as function of $w \in W$ becomes a morphism

$$l : W \to \mathcal{M}$$

where $\mathscr{M}$ is the opposite to the monoid of all functions $W^r \rightarrow W^r$.

Since the elements $\sigma_1, ..., \sigma_n$ generate W, it is clear that condition (1.2) is equivalent with either one of the conditions

(1.2″) $l(\sigma_i v, x) = l(\sigma_i, l(v, x))$

(1.2‴) $l(u\sigma_i, x) = l(u, l(\sigma_i, x))$

Therefore, the function l is completely determined by the functions

(1.3) $k_i : W^r \rightarrow W^r, \quad i = 1, ..., n$

(1.4) $k_i x = l(\sigma_i, x)$

Conversely, given functions k_i as in (1.3), there is a unique left action $l : W^{1+r} \rightarrow W^r$ for which (1.4) holds. We shall call l the *left exponential* of $(k_1, ..., k_n)$ and write $l = {}^s(k_1, ..., k_n)$.

The following two examples are of importance. The left exponential of the left successor functions, $L_i : W \rightarrow W$, is the function

$$L = {}^s(L_1, ..., L_n) : W^2 \rightarrow W$$
$$L(w, x) = wx$$

The left exponential of the right successor functions, $R_i : W \rightarrow W$, is the function

$$R = {}^s(R_1, ..., R_n) : W^2 \rightarrow W$$
$$R(w, x) = x(\rho w)$$

where $\rho : W \rightarrow W$ is the reversal function defined in I,5.

A *right action*

$$r : W^{s+1} \rightarrow W^s$$

is defined similarly. Formulas (1.1) and (1.2) are replaced by

(1.1)^r $r(x, 1) = x$

(1.2)^r $r(x, uv) = r(r(x, u), v)$

Formulas (1.2″) and (1.2‴) become

(1.2″)^r $r(x, \sigma_i v) = r(r(x, \sigma_i), v)$

(1.2‴)^r $r(x, u\sigma_i) = r(r(x, u), \sigma_i)$

Formula (1.4) becomes

$(1.4)^r$ $$k_i x = r(x, \sigma_i)$$

and r is called the *right exponential* of $(k_1, ..., k_n)$. Notation:
$r = (k_1, ..., k_n)^\S$.

The right and left exponentials of $(k_1, ..., k_n)$ are related by the
formulas

(1.5) $$r(x, w) = l(\rho w, x), \qquad l(w, x) = r(x, \rho w)$$

Consequently, the right exponentials of $(L_1, ..., L_n)$ and $(R_1, ..., R_n)$
are the functions

$$L' : W^2 \to W, \qquad R' : W^2 \to W$$
$$L'(x, w) = (\rho w) \, x, \qquad R'(x, w) = xw$$

In the case $n = 1$, $W = N$, the left exponential ${}^\S k$ of $k : N^r \to N^r$
is

$$ {}^\S k : N^{1+r} \to N^r $$
$$ {}^\S k(n, x) = k^n(x) $$

while the right exponential is

$$ k^\S : N^{r+1} \to N^r $$
$$ k^\S(x, n) = k^n(x) $$

In both cases $k^0(x) = x$ by convention.

2. Primitive Categories

An admissible subcategory $\mathscr{A}$ of $\mathscr{F}(W^{\boldsymbol{\cdot}})$ is called *primitive*
if it satisfies axiom (2.1), one of the two axioms (2.2) or (2.2'),
and one of the axioms (2.3) or (2.3') that follow.

(2.1)   *Unit.*   The function $U : W^{\boldsymbol{\cdot}} \to W$ with value 1 is in $\mathscr{A}$.

(2.2)   *Left successors.*   For each $1 \leqslant i \leqslant n$, the left successor
function $L_i : W \to W$ is in $\mathscr{A}$.

(2.2′)   *Right successors.*   For each $1 \leqslant i \leqslant n$, the right successor
function $R_i : W \to W$ is in $\mathscr{A}$.

(2.3)   *Left exponentiation.*   If $k_i : W^r \to W^r$, $1 \leqslant i \leqslant n$ are in $\mathscr{A}$,
then so is the left exponential ${}^\S(k_1 ,..., k_n) : W^{1+r} \to W^r$.

(2.3′)   *Right exponentiation.*   If $k_i : W^r \to W^r$, $1 \leqslant i \leqslant n$ are in $\mathscr{A}$,
then so is the right exponential $(k_1 ,..., k_n)^\S : W^{r+1} \to W^r$.

THEOREM 2.1.   *In a primitive category $\mathscr{A}$, all four axioms*
(2.2), (2.2′), (2.3), and (2.3′) hold. Further, the reversal function
$\rho : W \to W$ *is in $\mathscr{A}$.*

Proof.   In view of the formulas

$$R_i = \rho L_i \rho, \qquad L_i = \rho R_i \rho$$

and formulas (1.5) relating the right and left exponentials, it
suffices to show that ρ is in $\mathscr{A}$.

Assume that $\mathscr{A}$ satisfies (2.2′) and (2.3). The left exponential R
of $R_1 ,..., R_n$ satisfies $R(w, x) = x\rho w$. Thus $\rho w = R(w, 1)$ and

$$\rho : \xrightarrow{\; W \times U \;} W^2 \xrightarrow{\; R \;} W$$

Consequently, ρ is in $\mathscr{A}$.

Assume that $\mathscr{A}$ satisfies (2.2) and (2.3). The left exponential L
of $L_1 ,..., L_n$ satisfies $L(w, x) = wx$. Thus, $R_i w = L(w, \sigma_i)$ and

$$R_i : W \xrightarrow{\; W \times UL_i \;} W^2 \xrightarrow{\; L \;} W$$

Thus, R_i is in $\mathscr{A}$. This brings us back to the previous case.

The remaining two cases are handled symmetrically ∎

The class of all primitive subcategories of $\mathscr{F}(W^{\boldsymbol{\cdot}})$ is to be
denoted by $\mathbf{E}(W^{\boldsymbol{\cdot}})$. The smallest element of $\mathbf{E}(W^{\boldsymbol{\cdot}})$ is denoted by
$\mathscr{E}_0(W^{\boldsymbol{\cdot}})$. The functions in $\mathscr{E}_0(W^{\boldsymbol{\cdot}})$ are called *primitive recursive*
functions.

If Γ is any class of functions $f : W^r \to W^s$ (with various exponents r and s), we denote the smallest element of $\mathbf{E}(W^{\cdot})$ containing Γ by $\mathscr{E}_{\Gamma}(W^{\cdot})$. Thus, in particular, $\mathscr{E}_0(W^{\cdot}) = \mathscr{E}_{\phi}(W^{\cdot})$. The functions in $\mathscr{E}_{\Gamma}(W^{\cdot})$ are called Γ-*primitive* or *primitive recursive relative to* Γ.

EXERCISE 2.1.   *Show that all constant functions* $W^r \to W^s$ $(r \geqslant 0, s \geqslant 0)$ *are primitive recursive.*

EXERCISE 2.2.   *Show that if* $W \times f$ *is in* $\mathscr{A}$, *then* f *is in* $\mathscr{A}$.

3. Induction

Given functions

$$f : W^r \to W^s, \qquad h_i : W^{1+r+s} \to W^s, \qquad 1 \leqslant i \leqslant n$$

the function defined *inductively* from $(f, h_1, ..., h_n)$ is the function $g : W^{1+r} \to W^s$ satisfying

$$(3.1) \qquad g(1, x) = fx, \qquad g(\sigma_i w, x) = h_i(w, x, g(w, x))$$

for $w \in W$, $x \in W^r$, $1 \leqslant i \leqslant n$.

PROPOSITION 3.1.   *If* $\mathscr{A} \in \mathbf{E}(W^{\cdot})$ *and* $f, h_1, ..., h_n$ *are in* $\mathscr{A}$, *then so is* g, *where* g *is defined inductively from* $(f, h_1, ..., h_n)$.

Proof.   Let $k_i : W^{1+r+s} \to W^{1+r+s}$ be defined by

$$(3.2) \qquad k_i(w, x, y) = (\sigma_i w, x, h_i(w, x, y))$$

By two uses of Proposition I,4.3, k_i is in $\mathscr{A}$. Let $l = {}^{\S}(k_1, ..., k_n)$ be the left exponential of $k_1, ..., k_n$. We assert that

$$(3.3) \qquad l(w, 1, x, fx) = (w, x, g(w, x))$$

For $w = 1$, we have

$$l(1, 1, x, fx) = (1, x, fx) = (1, x, g(1, x))$$

as required. Assuming (3.3), we have

$$
\begin{aligned}
l(\sigma_i w, 1, x, fx) &= l(\sigma_i, l(w, 1, x, fx)) &&\text{by (1.2)} \\
&= k_i(l(w, 1, x, fx)) &&\text{by (1.4)} \\
&= k_i(w, x, g(w, x)) &&\text{by (3.3)} \\
&= (\sigma_i w, x, h_i(w, x, g(w, x))) &&\text{by (3.2)} \\
&= (\sigma_i w, x, g(\sigma_i w, x)) &&\text{by (3.1)}
\end{aligned}
$$

as required. This proves (3.3) by induction. Since l and f are in $\mathscr{A}$, it follows from (3.3) that g is in $\mathscr{A}$. Indeed, g is the composition

$$
W^{1+r} \xrightarrow{\ W \times U \times \langle W, f \rangle\ } W^{1+1+r+s} \xrightarrow{\ l\ } W^{1+r+s} \xrightarrow{\ \pi\ } W^s
$$

where $\pi(w, x, y) = y$ ∎

The induction procedure above may be called a "left" induction in that the left successor $\sigma_i w$ was used. We leave it to the reader to formulate and prove an appropriate "right" induction.

In the classical definition of "primitive recursive," induction (with $s = 1$) is called "primitive recursion" and is admitted as an axiom in place of our exponentiation axiom. One then proves that the class of "primitive recursive" functions so obtained is closed with respect to "simultaneous recursion" which is our induction (with $s \geqslant 0$). We now show that in the presence of the unit U and the left successors L_i, induction implies left exponentiation. This shows that our notion of "primitive recursive" coincides with the classical one.

Given $k_i : W^r \to W^r$, $1 \leqslant i \leqslant n$, in $\mathscr{A}$, define $f : W^r \to W^r$ and $h_i : W^{1+r+r} \to W^r$ by

$$
fx = x, \qquad h_i(w, x, y) = k_i y
$$

Let $g : W^{1+r} \to W^r$ be defined inductively from $f, h_1, \ldots, h_n$. Then g is in $\mathscr{A}$. Further,

$$
\begin{aligned}
g(1, x) &= fx = x \\
g(\sigma_i w, x) &= h_i(w, x, g(w, x)) \\
&= k_i(g(w, x))
\end{aligned}
$$

Thus, by (1. 2″) and (1.4), $g = {}^s(k_1, \ldots, k_n)$ ∎

More elaborate induction schemes will be discussed in Section 8.

EXERCISE 3.1. *Show that the predecessor function* $P : W \to W$, *defined by*

$$P1 = 1, \qquad P(\sigma_i w) = w$$

is primitive recursive.

4. The Case $W = N$

For the rest of this chapter, we shall be concerned with the case $W = N$; that is, the case when the alphabet Σ has only one letter σ_1. Some of the results are specific to this case (for example, results dealing with minimization (Section 6)). However, most of the significant results will later be extended to the general case. This will be achieved in Chapter III by means of a suitable coding (that is, a bijection) $W \to N$, using which results proved for N will yield analogous results for W.

As noted earlier, in dealing with a single letter alphabet, additive notation will be used and σ_1 and 1 will be replaced by 1 and 0. Thus, N will be the additive monoid of nonnegative integers. Because of commutativity, there is no difference between left and right successors. Further, there is only one successor function:

$$S : N \to N, \qquad Sx = 1 + x$$

The exponentiation takes a particularly simple form. The right exponential $l^s : N^{r+1} \to N^r$ of $l : N^r \to N^r$ is given by

$$l^s(x, y) = l^y x, \qquad x \in N^r, \quad y \in N$$

where l^0 is understood to be the identity function $N^r \to N^r$.

We list without proof the following primitive recursive functions $f : N^2 \to N$

$$f(x, y) = x + y$$
$$f(x, y) = xy$$
$$f(x, y) = x^y$$
$$f(x, y) = x \dot- y = \begin{cases} x - y & \text{if } x \geqslant y \\ 0 & \text{if } x < y \end{cases}$$
$$f(x, y) = |x - y| = (x \dot- y) + (y \dot- x)$$

A useful primitive recursive function $d : N^{t+t} \to N$ is the distance function

(4.1) $$d(x, y) = \sum_{i=1}^{t} \mid x_i - y_i \mid$$

for $x = (x_1 ,..., x_t)$, $y = (y_1 ,..., y_t)$.

THEOREM 4.1 (conditional definition). *Let*

$$g_0 , g_1 : N^r \to N^t$$
$$f_0 , f_1 : N^r \to N^s$$

be in $\mathscr{A} \in \mathbf{E}$. *Then the function*

$$f : N^r \to N^s$$

defined by

$$fx = \begin{cases} f_0 x & \text{if } g_0 x = g_1 x \\ f_1 x & \text{if } g_0 x \neq g_1 x \end{cases}$$

is in $\mathscr{A}$.

Proof. By virtue of Proposition I,4.3, we may assume $s = 1$. Consider the function $h : N^r \to N$ defined by $hx = e(d(g_0 x, g_1 x))$ where d is the distance function (4.1) while $e : N \to N$ is given by $ex = 1 \div x$. Then $hx = 1$ if $g_0 x = g_1 x$, and $hx = 0$ if $g_0 x \neq g_1 x$. Consequently,

$$f = h \cdot f_0 + (1 \div h) \cdot f_1$$

and thus f is in $\mathscr{A}$ ∎

5. Dominated Minimization

Given a function

$$f : N^{r+1} \to N$$

the *minimization*

$$\mu f : N^r \to N$$

is defined by setting

$$(\mu f)\, x = \inf\{t \mid f(x, t) = 0\}$$

If the bracketed set is empty, then $(\mu f)x$ is undefined; thus in general, μf is a partial function.

A function $f : N^r \to N$ will be called $\mathscr{A}$-*dominated* if there exists a function $g : N^r \to N$ in $\mathscr{A}$ such that $fx \leqslant gx$ for all $x \in N^r$.

THEOREM 5.1 (dominated minimization). *If* $f : N^{r+1} \to N$ *is in an* $\mathscr{A} \in \mathbf{E}$ *and* $\mu f : N^r \to N$ *is an* $\mathscr{A}$-*dominated function, then* μf *is in* $\mathscr{A}$.

Proof. Consider the primitive recursive function $e : N \to N$ defined by $e0 = 0$ and $ex = 1$ otherwise. Setting $f' = fe$, we have $\mu f' = \mu f$ and f' has only values 0 or 1. Further, f' is in $\mathscr{A}$.

Next replace f' by the function $f'' : N^{r+1} \to N$ given by

$$f''(x, y) = \prod_{t=0}^{y} f'(x, t)$$

Since

$$f''(x, 0) = f'(x, 0), \qquad f''(x, 1 + y) = f'(x, 1 + y)f''(x, y)$$

it follows by induction that f'' is in $\mathscr{A}$. Clearly, $\mu f'' = \mu f'$. The function f'' has only values 0 or 1 and further satisfies the condition

$$f''(x, 1 + y) \leqslant f''(x, y)$$

This implies

$$(\mu f)\, x = (\mu f'')\, x = \sum_{t=0}^{(\mu f)x} f''(x, t)$$

If $g : N^r \to N$ is a function in $\mathscr{A}$ such that $(\mu f)x \leqslant gx$, then we also have

$$(\mu f)\, x = \sum_{t=0}^{gx} f''(x, t)$$

because $f''(x, t) = 0$ for $t > (\mu f)x$. Next, define

$$h(x, y) = \sum_{t=0}^{y} f''(x, t)$$

Then h is in $\mathscr{A}$ and

$$(\mu f) x = h(x, gx)$$

Thus, μf is in $\mathscr{A}$ ∎

6. Primitive Recursive Sets

Given a subset A of N^r, the *characteristic* function

$$\chi A : N^r \to N$$

of A is defined by

$$(\chi A) x = \begin{cases} 1 & \text{if} \quad x \in A \\ 0 & \text{if} \quad x \in \bar{A} \end{cases}$$

where $\bar{A}$ denotes the complement $N^r \setminus A$ of A.

We note without proof the following formal properties of characteristic functions:

(6.1) $$\chi \bar{A} = 1 \dot{-} \chi A$$

(6.2) $$\chi(A \cap B) = (\chi A) \cdot (\chi B)$$

(6.3) $$\chi \varnothing = 0, \qquad \chi N^r = 1$$

(6.4)   If $f : N^s \to N^r$,   then $\chi(f^{-1}A) = f(\chi A)$

(6.5) $$\chi(A \times N) = (\chi A) \times N, \qquad \chi(N \times A) = N \times \chi A$$

Assume henceforth that $\mathscr{A} \in \mathbf{E}$. A subset A of N^r will be said to be in $\mathscr{A}$ if its characteristic function is in $\mathscr{A}$. If $\mathscr{A} = \mathscr{E}_0$, we say that A is *primitive recursive*.

From Properties (6.1)–(6.5) we obtain:

PROPOSITION 6.1. *If $A \subset N^r$ and $B \subset N^r$ are in $\mathscr{A}$, then so are $A \cup B$, $\bar{A}$ and $A \cap B$. The subsets $\varnothing$ and N^r of N^r are in $\mathscr{A}$.*

If $f : N^s \to N^r$ *is in* $\mathscr{A}$, *then so is the subset* $f^{-1}A$ *of* N^s. *If* $C \subset N^s$ *is in* $\mathscr{A}$, *then so is the subset* $A \times C$ *of* N^{r+s}.

For the last assertion, note that $A \times C = (N^r \times C) \cap (A \times N^s)$ ∎

PROPOSITION 6.2. *All finite subsets of* N^r *are in* $\mathscr{A}$.

Proof. In view of Proposition 6.1 it suffices to consider a single element x of N. For this we note

$$(\chi x)\, y = 1 \doteq |\, x - y \,| \quad ∎$$

PROPOSITION 6.3. *If* $f, g : N^r \to N$ *are in* $\mathscr{A}$, *then so is*

$$X = \{x \mid x \in N^r, \quad fx \leqslant gx\}$$
$$Y = \{x \mid x \in N^r, \quad fx = gx\}$$

Proof. $(\chi X)x = 1 \doteq (fx \doteq gx)$ and $(\chi Y)x = 1 \doteq |\, fx \doteq gx \,|$ ∎

COROLLARY 6.4. *If* $f, g : N^r \to N^s$ *are in* $\mathscr{A}$, *then the set*

$$X = \{x \mid x \in N^r, \quad fx = gx\}$$

is in $\mathscr{A}$ ∎

COROLLARY 6.5. *If* $f : N^r \to N^s$ *is in* $\mathscr{A}$, *then the graph* $\gamma f \subset N^{r+s}$ *is in* $\mathscr{A}$.

Proof. Indeed, $\gamma f = \{(x, y) \mid fx = y\}$ ∎

Given a subset X of N^{r+1}, the partial function

$$\mu X = \mu(\chi \bar{X}) : N^r \to N$$
$$(\mu X)\, x = \inf\{t \mid (x, t) \in X\}$$

is called the *minimization* of X. Theorem 5.1 yields:

THEOREM 6.6. *If* $X \subset N^{r+1}$ *is in* $\mathscr{A}$ *and if* μX *is an* $\mathscr{A}$-*dominated function, then* μX *is in* $\mathscr{A}$ ∎

Theorem 6.6 will now be used to establish the primitive recursiveness of some functions that will be needed in the sequel.

(6.6)   The "integral part" function

$$f : N^2 \to N, \qquad f(x, y) = \left[\frac{x}{y+1}\right]$$

is primitive recursive.

Indeed, consider the set

$$X = \{(x, y, z) \mid x < (z+1)(y+1)\} \subset N^3$$

which is primitive recursive by Proposition 6.3. Since $[x/y + 1]$ is the unique solution z of the inequality $z(y+1) \leqslant x < (z+1)(y+1)$, it follows that

$$f(x, y) = (\mu X)(x, y)$$

Since $f(x, y) \leqslant x$, it follows from Theorem 6.6 that f is primitive recursive ∎

(6.7)   For a fixed integer $n > 0$, the functions $q, r : N \to N$ defined by

$$q0 = r0 = 0$$

$$z = nqz + rz$$

$$1 \leqslant rz \leqslant n \text{ if } z > 0 \text{ are primitive recursive.}$$

Indeed,

$$qz = [z \div 1/n], \qquad rz = z \div nqz \quad \blacksquare$$

EXERCISE 6.1.   *Assume $f, g : N^r \to N$ are in $\mathscr{A}$ and that*

$$fx > 0, \qquad fx \text{ divides } gx$$

for all $x \in N^r$. Show that $g/f : N^r \to N$ is in $\mathscr{A}$.

EXERCISE 6.2.   *Show that the function $R : N^2 \to N$ defined by the conditions*

$$R(x, y) \equiv x \bmod y + 1, \qquad 0 \leqslant R(x, y) \leqslant y$$

is primitive recursive.

EXERCISE 6.3. *Show that the function* $\exp : N^2 \to N$

$$\exp(x, y) = \sup\{z \mid (y + 2)^z \text{ divides } x + 1\}$$

is primitive recursive.

7. Primitive Recursive Isomorphisms

We shall be interested here in establishing a bijection $b : N^2 \to N$ such that both b and b^{-1} are primitive recursive. Such a function will be called a *primitive recursive isomorphism*.

A bijection $N^2 \approx N$ is equivalent to an order of type ω in N^2. We arrive at such an order in the following way. As a first approximation, we order the pairs (x, y), with $x, y \in N$, by the number $t = x + y$. Then pairs (x, y), (x', y'), with $x + y = x' + y'$, are ordered by the first coordinate. This leads to the following two rules:

$$(x, 0) < (0, x + 1)$$
$$(x, y + 1) < (x + 1, y)$$

which define the order completely. The bijection $b : N^2 \to N$ that this order defines is easily proven by induction to be given by the formula

$$(7.1) \qquad b(x, y) = x + \sum_{i=0}^{x+y} i$$
$$= x + \tfrac{1}{2}(x + y)(x + y + 1)$$

which shows that b is primitive recursive. We note that

$$(7.2) \qquad\qquad\qquad b(0, 0) = 0$$

$$(7.3) \qquad\qquad b(x, y) \geqslant x, \qquad b(x, y) \geqslant y$$

To prove that $b^{-1} : N \to N^2$ is primitive recursive, we write $b^{-1} = \langle b', b'' \rangle$, where $b', b'' : N \to N$. Then by (7.3) we have $x = b(b'x, b''x) \geqslant b'x$. Thus,

$$(7.4) \qquad\qquad\qquad b'x \leqslant x, \qquad b''x \leqslant x$$

Consider the sets

$$X = \{(z, x, y) \mid z = b(x, y)\} \subset N^3$$

$$Y = \{(z, x) \mid (z, x, y) \in X \text{ for some } y \in N\} \subset N^2$$

We note that $(z, x, y) \in X$ if and only if $x = b'z$ and $y = b''z$. Thus, $Y = \gamma b'$ and therefore $b' = \mu Y$. Further, since $b''z \leqslant z$, we have

$$(\chi Y)(z, x) = 1 \dotdiv \prod_{y=0}^{z} (\chi \overline{X})(z, x, y)$$

Since by Proposition 6.3 the set X is primitive recursive, so is the function $\chi \overline{X}$. Thus, by the last formula, χY is primitive recursive and so is $b' = \mu Y$ in view of Theorem 6.6 and (7.4). The proof for b'' is similar ∎

Using $b : N^2 \to N$, we define bijections

$$b_r : N^r \to N, \qquad r > 0$$

as follows:

$$b_1 : N \to N \quad \text{is the identity}$$

$$b_{r+1} : N^{r+1} \xrightarrow{b_r \times N} N^2 \xrightarrow{b} N$$

EXERCISE 7.1. *Show that the function $2^x(1 + 2y) - 1$ is a primitive recursive isomorphism $N^2 \to N$.*

8. Various Induction Schemes

There is a variety of induction schemes more complicated than the one given in Section 3 under which primitive recursiveness is preserved. We shall give here a few rather simple ones that will be used in the sequel.

PROPOSITION 8.1. *Let*

$$f : N^{1+r} \to N^s, \qquad h : N^{2+r+s} \to N^s$$

be functions in $\mathscr{A} \in \mathbf{E}$. Then the function $g : N^{2+r} \to N^s$ defined by

$$g(0, y, z) = f(y, z)$$
$$g(x, y, z) = h(x, y, z, g(x - 1, y + 1, z)) \qquad \text{if} \quad x > 0$$

for $x, y \in N$, $z \in N^r$, is in $\mathscr{A}$.

  Proof. Define
$$l : N^{2+r+s} \to N^{2+r+s}$$
$$l(x, y, z, t) = (x + 1, y \dotminus 1, z, h(x + 1, y \dotminus 1, z, t))$$

Then for $x > 0$,

$$l(x - 1, y + 1, z, g(x - 1, y + 1, z))$$
$$= (x, y, z, h(x, y, g(x - 1, y + 1, z)))$$
$$= (x, y, z, g(x, y, z))$$

Iterating, we find for $k \leqslant x$,

$$l^k(x - k, y + k, z, g(x - k, y + k, z)) = (x, y, z, g(x, y, z))$$

so that for $k = x$,

$$l^x(0, x + y, z, f(x + y, z)) = (x, y, z, g(x, y, z))$$

Since ${}^s l$ and f are in $\mathscr{A}$, it follow that g is in $\mathscr{A}$ ∎

  PROPOSITION 8.2.   *Let*

$$f : N^r \to N^s, \qquad e : N^r \to N^r$$
$$h : N^{1+r+s} \to N^s$$

be functions in $\mathscr{A} \in \mathbf{E}$. Then the function $g : N^{1+r} \to N^s$, defined by

$$g(0, z) = fz$$
$$g(x, z) = h(x, z, g(x - 1, ez)) \qquad \text{if} \quad x > 0$$

is in $\mathscr{A}$.

Proof. Define

$$f' : N^{1+r} \to N^s, \qquad g' : N^{2+r} \to N^s$$
$$h' : N^{2+r+s} \to N^s$$

as follows:

$$f'(y, z) = f(e^y z)$$
$$g'(x, y, z) = g(x, e^y z)$$
$$h'(x, y, z, t) = h(x, e^y z, t)$$

Then

$$g'(0, y, z) = f'(y, z)$$
$$g'(x, y, z) = h'(x, y, z, g'(x - 1, y + 1, z)) \qquad \text{if} \quad x > 0$$

If f and h are in $\mathscr{A}$, then so are f' and h'. Therefore, g' is in $\mathscr{A}$ by Proposition 8.1. Since $g(x, z) = g'(x, 0, z)$, it follows that g is in $\mathscr{A}$ ∎

PROPOSITION 8.3. *Let*

$$f : N^r \to N^s, \qquad h : N^{r+s} \to N^s, \qquad k : N \to N$$

be functions in $\mathscr{A} \in \mathbf{E}$ and assume that

$$k0 = 0, \qquad kx < x \qquad \text{for} \quad x > 0$$

Then the unique function $g : N^{1+r} \to N^s$ satisfying

$$g(0, y) = fy$$
$$g(x, y) = h(y, g(kx, y)), \qquad x > 0$$

is in $\mathscr{A}$.

Proof. Let ex be the least exponent such that $k^{ex}x = 0$. Since $e = \mu k^s$ and $ex \leqslant x$, it follows from dominated minimization that $e : N \to N$ is in $\mathscr{A}$. Let $l : N^{r+s} \to N^{r+s}$ in $\mathscr{A}$ be given by

$$l(y, z) = (y, h(y, z))$$

We assert that

(8.1) $$l^{ex}(y, fy) = (y, g(x, y))$$

If $ex = 0$, then $x = 0$ and $g(x, y) = fy$ so that (8.1) holds. If $ex > 0$, then $x > 0$ and $ex = 1 + e(kx)$. Assuming that (8.1) holds with kx in place of x we have

$$l^{ex}(y, fy) = l(l^{e(kx)}(y, fy))$$
$$= l(y, g(kx, y))$$
$$= (y, h(y, g(kx, y)))$$
$$= (y, g(x, y))$$

as required by (8.1). Since e and l^s are in $\mathscr{A}$, it follows from (8.1) that g is in $\mathscr{A}$ ∎

It would appear from the above that any "reasonable" induction scheme leads from primitive recursive functions again to primitive recursive functions. This is not the case. The function $\Psi : N^2 \to N$, defined by the conditions

$$\Psi(0, y) = y + 1$$
$$\Psi(x + 1, 0) = \Psi(x, 1)$$
$$\Psi(x + 1, y + 1) = \Psi(x, \Psi(x + 1, y))$$

is not primitive recursive. This fact will be established in Appendix A.

Recursive Isomorphisms

1. Definitions and Results

We shall consider two finite nonempty alphabets Σ and Ω and we shall be interested in bijections

$$c : W \to V$$

where $W = \Sigma^*$, $V = \Omega^*$.

To simplify the notation, the function $c \times \cdots \times c : W^r \to V^r$ will simply be denoted by c. Similarly, $c^{-1} \times \cdots \times c^{-1} : V^r \to W^r$ will be denoted by c^{-1}.

Given any relation $f : W^r \to W^s$, we shall denote the composite relation

$$V^r \xrightarrow{c^{-1}} W^r \xrightarrow{f} W^s \xrightarrow{c} V^s$$

by f', which we shall call the *conjugate* of f. Similarly, if $g : V^r \to V^s$ is a relation, its *conjugate* $g' : W^r \to W^s$ is defined as $g' = cgc^{-1}$.

For any class $\mathscr{A}$ of relations in $\mathscr{R}(W^{\cdot})$, we denote by $\mathscr{A}'$ the class of all conjugates of the relations in $\mathscr{A}$.

PROPOSITION 1.1. *Conjugation establishes an isomorphism of categories* $\mathscr{R}(W^{\cdot}) \approx \mathscr{R}(V^{\cdot})$. *Under this isomorphism, admissible subcategories of* $\mathscr{R}(W^{\cdot})$ *are carried into admissible subcategories of* $\mathscr{R}(V^{\cdot})$ *and vice versa.*

The proof is entirely formal and is left to the reader ∎

With the above preliminaries, the main results may now be stated.

THEOREM 1.2. *For any bijection* $c : W \to V$, *the following conditions are equivalent.*

(i) *Conjugation by c establishes a bijection*

$$\mathbf{E}(W^{\cdot}) \approx \mathbf{E}(V^{\cdot})$$

(ii) $(\mathscr{E}_0(W^{\cdot}))' = \mathscr{E}_0(V^{\cdot})$.

(iii) *For each* $\sigma \in \Sigma$ *and each* $\omega \in \Omega$, *the funtions*

$$L_{\sigma}' = c^{-1}L_{\sigma}c : V \to V$$
$$L_{\omega}' = cL_{\omega}c^{-1} : W \to W$$

are primitive recursive.

If condition (ii) holds, we shall say that c is a *primitive recursive isomorphism*. Clearly, the inverse of a primitive recursive isomorphism is a primitive recursive isomorphism.

COROLLARY 1.3. *The composition of two primitive recursive isomorphisms is again a primitive recursive isomorphism* ∎

The implications (i) $\Rightarrow$ (ii) $\Rightarrow$ (iii) are clear.
In Section 2 we shall explicitly exhibit a bijection

$$a_W : W \to N$$

satisfying

$$a_W 1 = 0$$

Further, in Section 3 we shall show that a_W satisfies condition (i). *Taking this for granted*, we shall now complete the proof of the theorem above by proving the implication (iii) $\Rightarrow$ (i).
We need the following two lemmas.

LEMMA 1.4. *If $d : W \to W$ is a bijection and if $d^{-1}L_\sigma d$ is primitive recursive for each $\sigma \in \Sigma$, then d is primitive recursive.*

Proof. Set $h_\sigma = d^{-1}L_\sigma d$. Then

$$d(\sigma w) = d(L_\sigma w) = (L_\sigma d)\, w = (dd^{-1}L_\sigma d)\, w = (dh_\sigma)\, w = h_\sigma(dw)$$

so that d is primitive recursive by induction   ∎

LEMMA 1.5. *For every bijection $d : W \to W$, the following conditions are equivalent.*

(iv) *For every subcategory $\mathscr{A}$ of $\mathscr{R}(W^\cdot)$ containing $\mathscr{E}_0(W^\cdot)$,*

$$d^{-1}\mathscr{A}d = \mathscr{A}$$

(v) $d^{-1}\mathscr{E}_0(W^\cdot)d = \mathscr{E}_0(W^\cdot)$.

(vi) d *is a primitive recursive isomorphism.*

(vii) d *and d^{-1} are primitive recursive.*

Proof. The implication (iv) $\Rightarrow$ (v) is clear. From (v) we have $d\mathscr{E}_0(W^\cdot)\, d^{-1} = \mathscr{E}_0(W^\cdot)$ so that the implication (v) $\Rightarrow$ (vi) is clear. The implication (vi) $\Rightarrow$ (vii) follows from Lemma 1.4. Assuming (vii), we have $d^{-1}\mathscr{A}d \subset \mathscr{A}$ and $d\mathscr{A}\, d^{-1} \subset \mathscr{A}$. Hence, $\mathscr{A} \subset d^{-1}\mathscr{A}d$ so that (iv) follows   ∎

Let $a : W \to V$ be the composition $W \xrightarrow{a_W} N \xrightarrow{a_V^{-1}} V$. Since we are assuming that a_W and a_V satisfy condition (i), it follows that a satisfies condition (i) and, hence, also conditions (ii) and (iii).

We are now ready to prove the implication (iii) $\Rightarrow$ (i). Assume that $c : W \to V$ is a bijection satisfying (iii). Then $ac^{-1}L_\omega ca^{-1}$ is primitive recursive and by Lemma 1.4, so is ca^{-1}. Similarly, $a^{-1}cL_\omega c^{-1}a$ is primitive recursive so that $c^{-1}a$ is. Thus $a(c^{-1}a)\, a^{-1} = ac^{-1}$ is primitive recursive. Since ca^{-1} and its inverse are primitive recursive, we have by Lemma 1.5 that $ac^{-1}\mathscr{A}ca^{-1} = \mathscr{A}$ for every subcategory $\mathscr{A}$ of $\mathscr{R}(W^\cdot)$ containing $\mathscr{E}_0(W^\cdot)$. This is equivalent with

(1.1) $c^{-1}\mathscr{A}c = a^{-1}\mathscr{A}a$

Since (i) holds for $a : W \to V$, it also holds for $c : W \to V$ ∎

As a by-product of the arguments above, we obtain

PROPOSITION 1.6. *There exist primitive recursive isomorphisms*
$c : W \to V$ *satisfying* $c1 = 1$. *The bijection* $\mathbf{E}(W^{\cdot}) \approx \mathbf{E}(V^{\cdot})$ *of* (i)
is independent of c ∎

PROPOSITION 1.7. *A primitive recursive isomorphism* $c : W \to V$
establishes a bijection between all subcategories $\mathscr{A}$ *of* $\mathscr{R}(W^{\cdot})$ *containing*
$\mathscr{E}_0(W^{\cdot})$ *and all subcategories* $\mathscr{A}'$ *of* $\mathscr{R}(V^{\cdot})$ *containing* $\mathscr{E}_0(V^{\cdot})$. *This*
bijection is independent of the choice of c ∎

EXERCISE 1.1. *Prove that a bijection* $c : W \to V$ *is a primitive*
recursive isomorphism if and only if the conjugates of the monoid mul-
tiplications $W^2 \to W$ *and* $V^2 \to V$ *are primitive recursive.*

2. Description of the Bijection $a : W \to N$

A bijection $W \approx N$ is equivalent to an order of type ω in W.
To achieve this, we use the length of a word in W as a first
approximation. Then elements of the same length are ordered
lexicographically with precedence given to the last letter. Thus the
order is defined by the following conditions

$$|w'| < |w| \Rightarrow w' < w$$
$$|w'| = |w| \ \& \ 1 \leqslant i < j \leqslant n \Rightarrow w'\sigma_i < w\sigma_j$$
$$|w'| = |w| \ \& \ w' < w \Rightarrow w'\sigma_i < w\sigma_i$$

The resulting enumeration of W begins as follows:

$$1, \sigma_1 ,..., \sigma_n , \sigma_1\sigma_1 ,..., \sigma_n\sigma_1 , \sigma_1\sigma_2 ,..., \sigma_n\sigma_2 ,...$$

This leads to the following closed formula for $a : W \to N$.

$$(2.1) \qquad\qquad aw = \sum_{j=1}^{l} i_j n^{j-1} \qquad \text{for} \quad w = \sigma_{i_1} \cdots \sigma_{i_l}$$

The bijection a may be also described inductively

$$(2.2) \qquad\qquad a1 = 0, \qquad a(\sigma_i v) = i + nav$$

PROPOSITION 2.1.   *The composition*

$$S' : W \xrightarrow{a} N \xrightarrow{S} N \xrightarrow{a^{-1}} W$$

is primitive recursive.

Proof.   Indeed, one readily verifies

$$S'(\sigma_n w) = L_1(S'w)$$
$$S'(\sigma_i w) = L_{i+1}s, \qquad \text{for} \quad 1 \leqslant i < n$$

Thus, S' is primitive recursive by induction   ∎

PROPOSITION 2.2.   *The compositions*

$$L_i' : N \xrightarrow{a^{-1}} W \xrightarrow{L_i} W \xrightarrow{a} N$$

for $1 \leqslant i \leqslant n$ are primitive recursive.

Proof.   Setting $x = av$, we have $a^{-1}x = v$, $L_i(a^{-1}x) = L_i v$, $L_i'x = a(L_i(a^{-1}x)) = a(L_i v) = a(\sigma_i v)$ so that by (2.2)

$$L_i'x = i + nx \quad ∎$$

The two last propositions show that a satisfies condition (iii) of Theorem 1.2. To complete the program laid down in Section 1, we must prove that a satisfies condition (i) of Theorem 1.2. This will be done in the next section.

3. Proof of Condition (i)

Let $\mathscr{A} \in \mathbf{E}(W^*)$. It must be shown that $\mathscr{A}' \in \mathbf{E}(N^*)$. The conjugate of the unit $U_W : W^0 \to W$ is the unit $U_N : N^0 \to N$. Thus U_N is in $\mathscr{A}'$. The conjugate S' of the successor $S : N \to N$ is primitive recursive by Proposition 2.1. Hence, S' is in $\mathscr{A}$, so that S is in $\mathscr{A}'$. Thus, only the exponentiation axiom in $\mathscr{A}'$ needs to be verified.

Let, then, $f : W^r \to W^r$ and define

$$h : W^{r+1} \to W^r, \qquad h(x, w) = f^{aw}x$$

Then, for $y \in N^r$, $z \in N$, we have

$$h'(y, z) = a(h(a^{-1}y, a^{-1}z))$$
$$= a(f^z(a^{-1}y))$$
$$= (f^z)' y = f'^z y = f'^s(y, z)$$

Thus h' is the exponential of f' and it suffices to show that h is in $\mathscr{A}$.

Choose $k_1 = f$ and let $k_2 ,..., k_n : W^r \to W^r$ be arbitrary functions in $\mathscr{A}$. Let $l = (k_1 ,..., k_n)^s$.

The equalities

$$l(x, 1) = x, \qquad l(x, w\sigma_1) = f(l(x, w))$$

imply

$$l(x, \sigma_1^n) = f^n x, \qquad n \geqslant 0$$

Hence

$$h(x, w) = f^{aw}x = l(x, \sigma_1^{aw})$$

Since l is in $\mathscr{A}$, it suffices, therefore, to show that the function

$$e : W \to W, \qquad ew = \sigma_1^{aw}$$

is primitive recursive. From the inductive description of $a : W \to N$ given in (2.2), we deduce

$$e(\sigma_i w) = \sigma_1^{a\sigma_i w} = \sigma_1^{i+naw} = \sigma_1^i \sigma_1^{naw} = \sigma_1^i(\sigma_1^{aw})^n = \sigma_1^i(ew)^n = m_i(ew)$$

where

$$m_i : W \to W$$

is defined by

$$m_i w = \sigma_1^i w^n$$

Since m_i is clearly primitive recursive, it follows by induction that e is primitive recursive.

Next consider $\mathscr{A} \in \mathbf{E}(N^*)$. It is required to show that $\mathscr{A}' \in \mathbf{E}(W^*)$. The unit $U_W : W^0 \to W$ is in $\mathscr{A}'$ since it is the conjugate of $U_N : N^0 \to N$. By Proposition 2.2, the conjugate $L_i' : N \to N$ of the successor function $L_i : W \to W$ is primitive recursive so that

L_i' is in $\mathscr{A}$. Thus, L_i is in $\mathscr{A}'$. It remains only to verify the exponentiation axiom in $\mathscr{A}'$.

Let then $k_1, ..., k_n : N^r \to N^r$ be functions in $\mathscr{A}$ and let $l' = (k_1', ..., k_n')^s : W^{r+1} \to W^r$. It is required to show that $l : N^{r+1} \to N^r$ is in $\mathscr{A}$. It will be convenient to introduce the function $k_0 : N^r \to N^r$ defined by $k_0 y = y$ for $y \in N^r$.

The proof utilizes the primitive recursive functions $q : N \to N$, $r : N \to N$ of II(7.7), defined by

$$q0 = 0 = r0$$

$$z = rz + nqz, \qquad 1 \leqslant rz \leqslant n \qquad \text{if} \quad z > 0$$

The definition above compared with formula (2.2) yields

(3.1) $$r(a(\sigma_i v)) = i, \qquad q(a(\sigma_i v)) = av$$

We next introduce the following functions

$$\begin{aligned}
f &: N \to N, & fz &= |a^{-1}z| \\
g &: N^{r+1} \to N^r, & g(y, z) &= k_{rz} y \\
h &: N^{r+1} \to N^{r+1}, & h(y, z) &= (g(y, z), qz)
\end{aligned}$$

and claim that

(3.2) $$(l(y, z), 0) = h^s(y, z, fz) = h^{fz}(y, z)$$

To verify the claim, take $y = ax$, $z = aw$ where $x \in W^r$, $w \in W$. Then (3.2) becomes

(3.3) $$(l'(x, w), 1) = h'^{|w|}(x, w)$$

We prove (3.3) by induction on the length of w. If $w = 1$, both sides yield $(x, 1)$. Now assume $w = \sigma_i v$ and assume that (3.3) holds with v in place of w. Then

$$\begin{aligned}
h'(x, \sigma_i v) &= a^{-1}(h(ax, a(\sigma_i v))) \\
&= a^{-1}(k_{r(a, (\sigma_i v))}(ax), q(a(\sigma_i v))) \\
&= a^{-1}(k_i(ax), av) & \text{by (3.1)} \\
&= (k_i' x, v)
\end{aligned}$$

Consequently,

$$
\begin{aligned}
h'^{|w|}(x, w) &= h'^{|v|}(h'(x, \sigma_i v)) \\
&= h'^{|v|}(k_i'x, v) \\
&= (l'(k_i'x, v), 1) \qquad \text{by (3.3) for } v \\
&= (l'(x, \sigma_i v), 1) \\
&= (l'(x, w), 1)
\end{aligned}
$$

This proves (3.3) and with it the claim (3.2).

To conclude that l is in $\mathscr{A}$, it suffices to show that f is primitive recursive and g is in $\mathscr{A}$.

To prove that g is in $\mathscr{A}$, consider the functions

$$
c : N^{r(1+n)} \to N^{r(1+n)}
$$
$$
\pi : N^{r(1+n)} \to N^r
$$
$$
c(x_0, ..., x_n) = (x_1, ..., x_n, x_0)
$$
$$
\pi(x_0, ..., x_n) = x_0 \qquad \text{for } x_0, x_1, ..., x_n \in N^r
$$

These functions are primitive recursive and, therefore, so is the function

$$
d = c^s \pi : N^{r(1+n)+1} \to N^r
$$
$$
d(x_0, ..., x_n, t) = x_i \qquad \text{if} \qquad t \equiv i \bmod(1 + n)
$$

Since

$$
g(y, z) = k_{rz} y = d(k_0 y, ..., k_n y, rz)
$$

it follows that g is in $\mathscr{A}$.

Clearly, $f0 = 0$. Assume $z > 0$. Then $|a^{-1}z| > 0$. Let $a^{-1}z = \sigma_i v$. Then by (3.1),

$$
\begin{aligned}
fz &= |\sigma_i v| \\
&= 1 + |v| \\
&= 1 + f(a(v)) \\
&= 1 + f(q(a(\sigma_i v))) \\
&= 1 + f(qz)
\end{aligned}
$$

Thus Proposition II, 8.3 shows that f is primitive recursive ∎

4. Translated Results

We are now in a position to translate some earlier results concerning N to results concerning W. Let $\mathscr{A} \in \mathbf{E}(W^{\cdot})$. In order to give meaning to the statement "$A \subset W^r$ is in $\mathscr{A}$," we define the characteristic function

$$\chi A = (\chi A')' : W^r \to W$$

where A' is the image of A under $a : W \to N$. Thus, since $a\sigma_1 = 1 \in N$,

$$(\chi A)\, x = \begin{cases} \sigma_1 & \text{if } x \in A \\ 1 & \text{if } x \in \bar{A} \end{cases}$$

With this definition, we may now state some generalizations of earlier results.

PROPOSITION 4.1.   *If $A \subset W^r$, $B \subset W^r$ are in $\mathscr{A}$ then so are $A \cup B$, $\bar{A}$ and $A \cap B$. The subsets ϕ and W^r of W^r are in $\mathscr{A}$. If $f : W^s \to W^r$ is in $\mathscr{A}$, then so is the subset $f^{-1}A$ of W^s. If $C \subset W^s$ is in $\mathscr{A}$, then so is the subset $A \times C$ of W^{r+s}* ∎

PROPOSITION 4.2.   *All finite subsets of W^r are in $\mathscr{A}$* ∎

THEOREM 4.3.   *The bijection $b' = aba^{-1} : W^2 \to W$ is a primitive recursive isomorphism* ∎

CHAPTER IV

Recursive Relations

1. Distinguished Subcategories of $\mathscr{R}(W^{\cdot})$

Given a family of relations $f_i : X \to Y$, $i \in I$, we denote by $f = \bigcup f_i$ the relation $f : X \to Y$ defined by $fx = \bigcup f_i x$ for $x \in X$. Given a relation $f : X \to X$, the *closure* of f is the relation

$$f^+ : X \to X, \qquad f^+ = f \cup f^2 \cup \cdots \cup f^n \cup \cdots$$

A subcategory $\mathscr{A}$ of $\mathscr{R}(W^{\cdot})$ is called *distinguished* if it is admissible and if it satisfies the following axioms.

(1.1) *Unit.* The function $U : W^0 \to W$ is in $\mathscr{A}$.

(1.2) *Successors.* The functions $L_i : W \to W$ are in $\mathscr{A}$ for $1 \leqslant i \leqslant n$.

(1.3) *Union.* If $f, g : W^r \to W^s$ are in $\mathscr{A}$, then so is

$$f \cup g : W^r \to W^s$$

(1.4) *Closure.* If $f : W^r \to W^r$ is in $\mathscr{A}$, then so is $f^+ : W^r \to W^r$.

(1.5) *Inverse.* If $f : W^r \to W^s$ is in $\mathscr{A}$, then so is $f^{-1} : W^s \to W^r$.

The class of all distinguished subcategories of $\mathscr{R}(W^{\cdot})$ will be denoted by $\mathbf{R}(W^{\cdot})$ or $\mathbf{R}$ for brevity. *In what follows, it will be assumed that $\mathscr{A} \in \mathbf{R}$.*

PROPOSITION 1.1. *If $f, g : W^r \to W^s$ are in $\mathscr{A}$, then so is the relation $f \cap g : W^r \to W^s$ defined by $(f \cap g)x = fx \cap gx$.*

Proof. Note that $f \cap g$ is the composition

$$W^r \xrightarrow{\langle f, g \rangle} W^{s+s} \xrightarrow{\varDelta^{-1}} W^s$$

where $\varDelta x = (x, x)$ is the diagonal mapping ∎

PROPOSITION 1.2. *All constant functions $f : W^r \to W^s$ are in $\mathscr{A}$.*

Proof. In view of (1.2), it suffices to consider the constant function with value 1. This function is the composition

$$W^r \xrightarrow{\pi} W^0 \xrightarrow{U^s} W^s$$

and thus is in $\mathscr{A}$ ∎

PROPOSITION 1.3. *All empty relations $f : W^r \to W^s$ are in $\mathscr{A}$.*

Proof. We have $f = gh^{-1}$ where $g : W^r \to W$, $h : W^s \to W$ are constant relations with different constants ∎

Given a subset A of W^r, we denote by

$$\tau A : W^r \to W^0$$

the relation with domain A.

PROPOSITION 1.4. *A relation $f : W^r \to W^s$ is in $\mathscr{A}$ if and only if the relation $\tau \gamma f : W^{r+s} \to W^0$ is in $\mathscr{A}$.*

Proof. Note that $\tau \gamma f$ is the composition

$$W^{r+s} \xrightarrow{\langle W^r, f \rangle^{-1}} W^r \xrightarrow{\pi} W^0$$

where π is the unique function. Thus if f is in $\mathscr{A}$, then so is $\tau \gamma f$.

To prove the opposite implication, note that f is the composition

$$W^r \xrightarrow{\pi_1^{-1}} W^{r+s} \xrightarrow{\langle \tau \gamma f, W^{r+s} \rangle} W^{r+s} \xrightarrow{\pi_2} W^s$$

where π_1 and π_2 are projections   ∎

Note that the results so far used neither axiom (1.3) nor (1.4).

PROPOSITION 1.5. *In the presence of the other axioms, axioms* (1.3) *and* (1.4) *may be replaced by the single axiom*

(1.4*)   If $f : W^r \to W^r$ is in $\mathscr{A}$, then so is the relation $f^* : W^r \to W^r$, defined by

$$f^* = W^r \cup f \cup f^2 \cup \cdots \cup f^n \cup \cdots$$

Proof. Since

$$f^* = W^r \cup f^+$$

it follows that (1.3) and (1.4) imply (1.4*). Now assume (1.4*). Since

$$f^+ = ff^*$$

axiom (1.4) holds. To prove (1.3), consider $f, g : W^r \to W^s$ in $\mathscr{A}$. By Proposition 1.4, we may limit our attention to the case $s = 0$. If the relation g is empty, there is nothing to prove since $f \cup g = f$. Let then $x \in \operatorname{Dom} g$ and let $h : W^0 \to W^r$ be the function with value x. Then h is in $\mathscr{A}$ by Proposition 1.2. Further,

$$fhg = f, \qquad fhf \subset f$$

The inclusion implies $(fh)^+ = fh$. Therefore,

$$(fh)^* g = (W^r \cup fh) g = g \cup fhg = g \cup f \quad ∎$$

Remark. The proofs above do not fully use the algebraic structure of W. With appropriate changes, they remain valid for an admissible subcategory $\mathscr{A}$ of $\mathscr{R}(X^{\cdot})$, X any set, with proper axioms postulated on $\mathscr{A}$. In fact, Proposition 1.1 requires no axioms. Proposition 1.2 requires the assumption that all functions $X^0 \to X$ are in $\mathscr{A}$. Proposition 1.3 requires the further assumption that card $X > 1$. Proposition 1.4 requires the inversion axiom (1.5) while Proposition 1.5 utilizes only Propositions 1.2 and 1.4.

2. Further Consequences of the Axioms

THEOREM 2.1. *If $\mathscr{A} \in \mathbf{R}$, then $\mathscr{A} \cap \mathscr{F} \in \mathbf{E}$.*

Proof. Let $\mathscr{B} = \mathscr{A} \cap \mathscr{F}$. Then $\mathscr{B}$ is an admissible subcategory of $\mathscr{F}$ containing U and the successors L_i. We show that $\mathscr{B}$ is closed under right exponentiation. Let then $k_1, ..., k_n : W^r \to W^r$ be in $\mathscr{B}$ and $l : W^{r+1} \to W^r$ be the right exponential of $(k_1, ..., k_n)$. Define

$$g = \bigcup k_i \times L_i^{-1} : W^{r+1} \to W^{r+1}, \quad 1 \leqslant i \leqslant n$$
$$f = W^r \times U^{-1} : W^{r+1} \to W^r$$

Then

$$g(x, \sigma_i w) = (k_i x, w)$$
$$g(x, 1) = \varnothing$$
$$f(x, w) = \varnothing \quad \text{if} \quad w \neq 1$$
$$f(x, 1) = x$$

By the axioms and by Proposition 1.5, g and f are in $\mathscr{A}$. To conclude the proof, it thus suffices to show that $l = g * f$. This follows from the following assertion:

$$(g^n f)(x, w) = l(x, w) \quad \text{if} \quad n = |w|$$
$$= \varnothing \quad \text{otherwise}$$

which we verify by induction on n. For $n = 0$, we have $f(x, 1) = x = l(x, 1)$ and $f(x, w) = \varnothing$ if $w \neq 1$. Thus the assertion holds. Next, observe

$$(g^{n+1} f)(x, w) = (g^n f)[g(x, w)]$$

Thus, if $w = 1$, then $g(x, w) = \varnothing$ and $(g^{n+1}f)(x, w) = \varnothing$. If $w = \sigma_i v$, then $g(x, w) = (k_i x, v)$ and thus

$$(g^{n+1}f)(x, w) = (g^n f)(k_i x, v)$$

If $n + 1 \neq |w|$, then $n \neq |v|$ and the result is $\varnothing$. If $n + 1 = |w|$, then $n = |v|$ and by the inductive hypothesis we have

$$(g^n f)(k_i x, v) = l(k_i x, v) = l(l(x, \sigma_i), v)$$
$$= l(x, \sigma_i v) = l(x, w)$$

as required ∎

COROLLARY 2.2. *If $\mathscr{A} \in \mathbf{R}$, then $\mathscr{E}_0 \subset \mathscr{A}$; that is, all primitive recursive functions are in $\mathscr{A}$* ∎

THEOREM 2.3. *For every primitive recursive isomorphism $c : W \to V$, conjugation by c establishes a bijection*

$$\mathbf{R}(W^*) \approx \mathbf{R}(V^*)$$

which is independent of the choice of c.

Proof. This follows from Proposition III,1.7, Corollary 2.2, and the remark that axioms (1.3)–(1.5) are invariant under conjugation ∎

EXERCISE 2.1. *Call a bijection $c : W \to V$ recursive if for each $\sigma \in \Sigma$ the function*

$$L_\sigma' = c^{-1} L_\sigma c : V \to V$$

is recursive. Show that $c^{-1} : V \to W$ also is recursive. Establish the analog of Proposition III,1.7 with $\mathscr{E}_0$ replaced by $\mathscr{F}_0$. Prove the analog of Exercise III,1.1.

3. The Main Theorem

Let Γ be a class of functions $f : W^r \to W^s$. In Chapter II, Section 2 we denoted the smallest element of $\mathbf{E}$ containing Γ by $\mathscr{E}_\Gamma$.

Similarly we shall denote by $\mathcal{R}_\Gamma$ the smallest element of **R** containing Γ. We further introduce

$$\mathcal{P}_\Gamma = \mathcal{R}_\Gamma \cap \mathcal{P}, \qquad \mathcal{F}_\Gamma = \mathcal{R}_\Gamma \cap \mathcal{F}$$

The relations in $\mathcal{R}_\Gamma$ are called Γ-*recursive*. Thus $\mathcal{P}_\Gamma$ consists of the Γ-recursive partial functions and $\mathcal{F}_\Gamma$ consists of the Γ-recursive functions.

If $\Gamma = \varnothing$, we write $\mathcal{R}_0$, $\mathcal{P}_0$, and $\mathcal{F}_0$ instead of $\mathcal{R}_\varnothing$, $\mathcal{P}_\varnothing$, and $\mathcal{F}_\varnothing$. Further, we replace "$\varnothing$-recursive" by "recursive."

There is a deep connection between the categories $\mathcal{E}_\Gamma$ and $\mathcal{R}_\Gamma$ that will be stated below. The following definition is needed.

Given any subcategory $\mathcal{A}$ of $\mathcal{R}(W^{\cdot})$, we denote by $\mathcal{A}^{-1}\mathcal{A}$ the class of all relations

$$f : W^r \to W^s$$

such that either f is empty or there exist relations

$$W^r \xleftarrow{g} W^t \xrightarrow{h} W^s$$

in $\mathcal{A}$ such that $f = g^{-1}h$. Since we may take $t = r$ and $g = W^r$, it follows that

$$\mathcal{A} \subset \mathcal{A}^{-1}\mathcal{A}$$

THEOREM 3.1 (main theorem). *If $\mathcal{A} \in \mathbf{E}$, then $\mathcal{A}^{-1}\mathcal{A} \in \mathbf{R}$.*

The proof will be given in the next section.

COROLLARY 3.2.

$$\mathcal{E}_\Gamma^{-1}\mathcal{E}_\Gamma = \mathcal{F}_\Gamma^{-1}\mathcal{F}_\Gamma = \mathcal{P}_\Gamma^{-1}\mathcal{P}_\Gamma = \mathcal{R}_\Gamma^{-1}\mathcal{R}_\Gamma = \mathcal{R}_\Gamma$$

Proof. Indeed, the inclusions $\mathcal{E}_\Gamma \subset \mathcal{F}_\Gamma \subset \mathcal{P}_\Gamma \subset \mathcal{R}_\Gamma$ imply

$$\mathcal{E}_\Gamma^{-1}\mathcal{E}_\Gamma \subset \mathcal{F}_\Gamma^{-1}\mathcal{F}_\Gamma \subset \mathcal{P}_\Gamma^{-1}\mathcal{P}_\Gamma \subset \mathcal{R}_\Gamma^{-1}\mathcal{R}_\Gamma$$

The inclusion $\mathcal{R}_\Gamma^{-1}\mathcal{R}_\Gamma \subset \mathcal{R}_\Gamma$ follows from Proposition 1.3 and axiom (1.4). Finally, the inclusion $\mathcal{R}_\Gamma \subset \mathcal{E}_\Gamma^{-1}\mathcal{E}_\Gamma$ follows from $\Gamma \subset \mathcal{E}_\Gamma \subset \mathcal{R}_\Gamma$ and the main theorem ∎

COROLLARY 3.3. *Every nonempty relation* $f : W^r \to W^s$ *in* $\mathscr{R}_\Gamma$ *has the form* $f = g^{-1}h$, *where*

$$g : W \to W^r, \qquad h : W \to W^s$$

are functions in $\mathscr{E}_\Gamma$.

Proof. Indeed, by the above, we have $f = g^{-1}h$ with

$$g : W^t \to W^r, \qquad h : W^t \to W^s$$

in $\mathscr{E}_\Gamma$. If $t = 0$, then replace g and h by the compositions

$$W \xrightarrow{\pi} W^0 \xrightarrow{g} W^r, \qquad W \xrightarrow{\pi} W^0 \xrightarrow{h} W^s$$

where π is the unique function. If $t > 1$, then replace g and h by the compositions

$$W \xrightarrow{l^{-1}} W^t \xrightarrow{g} W^r, \qquad W \xrightarrow{l^{-1}} W^t \xrightarrow{h} W^s$$

where $l : W^t \to W$ is the bijection obtained by conjugation from the bijections $b_t : N^t \to N$ of II,7 ∎

EXERCISE 3.1. *Show that the condition* $f = g^{-1}h$ *is equivalent to*

$$\gamma f = \mathrm{Im}\langle g, h \rangle$$

(For any relation $k : W^r \to W^s$, the *image* $\mathrm{Im}\, k$ is defined as $\mathrm{Dom}\, k^{-1}$.)

4. Proof of the Main Theorem

We must verify that if $\mathscr{A} \in \mathbf{E}$, then $\mathscr{A}^{-1}\mathscr{A}$ is an admissible subcategory of $\mathscr{R}$ satisfying axioms (1.1)–(1.4). The difficult part of the argument is the verification of the closure axiom (1.3). The rest of the reasoning is fairly simple and makes use of very few properties of $\mathscr{A}$.

Since $\mathscr{A} \subset \mathscr{A}^{-1}\mathscr{A}$, it follows that $\mathscr{A}^{-1}\mathscr{A}$ contains the identity function, logical functions, unit and successors. Also, the inversion axiom (1.4) clearly holds in $\mathscr{A}^{-1}\mathscr{A}$. That $\mathscr{A}^{-1}\mathscr{A}$ is closed under cylindrification follows from the formula

$$W \times g^{-1}h = (W \times g)^{-1}(W \times h)$$

Next we show that $\mathscr{A}^{-1}\mathscr{A}$ is closed under composition. Consider functions

$$W^r \xleftarrow{g_1} W^t \xrightarrow{h_1} W^s$$

$$W^s \xleftarrow{g_2} W^u \xrightarrow{h_2} W^v$$

in $\mathscr{A}$ and let $f_1 = g_1^{-1}h_1$, $f_2 = g_2^{-1}h_2$. We may assume that f_1f_2 is nonempty. Then there exists $x_0 \in W^t$, $y_0 \in W^u$ with $h_1x_0 = g_2y_0$. Define

$$g: W^{t+u} \to W^r, \qquad h: W^{t+u} \to W^v$$

$$g(x, y) = \begin{cases} g_1x & \text{if } h_1x = g_2y \\ g_1x_0 & \text{otherwise} \end{cases}$$

$$h(x,y) = \begin{cases} h_2y & \text{if } h_1x = g_2y \\ h_2y_0 & \text{otherwise} \end{cases}$$

Then g and h are in $\mathscr{A}$ by the conditional definition theorem. The verification that $f_1f_2 = g^{-1}h$ is straightforward.

Next we prove that $\mathscr{A}^{-1}\mathscr{A}$ is closed under union. Let then

$$W^r \xleftarrow{g_i} W^{t_i} \xrightarrow{h_i} W^s, \qquad i = 1, 2$$

be in $\mathscr{A}$. Define

$$g: W^{t_1+t_2} \to W^r, \qquad h: W^{t_1+t_2} \to W^s$$

$$g(x, y) = \begin{cases} g_1x & \text{if } y = 1 \\ g_2x & \text{if } y \neq 1 \end{cases}$$

$$h(x, y) = \begin{cases} h_1x & \text{if } y = 1 \\ h_2x & \text{if } y \neq 1 \end{cases}$$

Then $g^{-1}h = g_1^{-1}h_1 \cup g_2^{-1}h_2$.

This brings us to the proof of the closure axiom. Here, it is convenient to apply conjugation and thus conduct the proof only for the case $W = N$.

Let then $f : N^r \to N^r$ be in $\mathscr{A}^{-1}\mathscr{A}$. We shall prove that f^+ also is in $\mathscr{A}^{-1}\mathscr{A}$.

If f is empty, there is nothing to prove. Thus we may assume $f = g^{-1}h$ with

$$N^r \xleftarrow{\;g\;} N^t \xrightarrow{\;h\;} N^r$$

in $\mathscr{E}_r$. Further, by Corollary 3.3, we may assume $t = 1$.

The proof will make heavy use of the bijection $b : N^2 \to N$. It will be convenient to write

$$b^{-1}x = (cx, dx)$$

Define

$$G, H : N^2 \to N^r$$

as follows:

$$G(0, z) = gz$$
$$G(x + 1, z) = \begin{cases} g(dz) & \text{if} \quad G(x, cz) = h(dz) \\ g0 & \text{if} \quad G(x, cz) \neq h(dz) \end{cases}$$
$$H(0, z) = hz,$$
$$H(x + 1, z) = \begin{cases} H(x, cz) & \text{if} \quad G(x, cz) = h(dz) \\ h0 & \text{if} \quad G(x, cz) \neq h(dz) \end{cases}$$

We first prove that G and H are in $\mathscr{A}$. For this we define

$$l : N^{1+2r} \to N^{2r}$$
$$l(z, u, v) = \begin{cases} (g(dz), v) & \text{if} \quad u = h(dz) \\ (g0, h0) & \text{otherwise} \end{cases}$$
$$K = \langle G, H \rangle : N^2 \to N^{2r}$$
$$K(x, z) = (G(x, z), H(x, z))$$

Assume that $x > 0$. If $G(x - 1, cz) = h(dz)$, then

$$\begin{aligned} K(x, z) &= (G(x, z), H(x, z)) \\ &= (g(dz), H(x - 1, cz)) \\ &= l(z, G(x - 1, cz), H(x - 1, cz)) \\ &= l(z, K(x - 1, cz)) \end{aligned}$$

If $G(x - 1, cz) \neq h(dz)$, then

$$
\begin{aligned}
K(x, z) &= (G(x, z), H(x, z)) = (g0, h0) \\
&= l(z, G(x - 1, cz), H(x - 1, cz)) \\
&= l(z, K(x - 1, cz))
\end{aligned}
$$

Thus in all cases, we have

$$K(x, z) = l(z, K(x - 1, cz)) \qquad \text{if} \quad x > 0$$

Further

$$K(0, z) = (gz, hz)$$

Since g, h, c, and d are in $\mathscr{A}$, it follows that l is in $\mathscr{A}$ by conditional definition. The fact that K is in $\mathscr{A}$ now follows from Proposition II,8.2. Thus G and H are in $\mathscr{A}$.

We assert that

(4.1) $$f^+ = G^{-1}H$$

We first show that

(4.2) $$f^t \subset G^{-1}H \qquad \text{for} \quad t > 0$$

We proceed by induction with respect to t. If $t = 1$, then $f = g^{-1}h \subset G^{-1}H$ since $G(0, z) = gz$, $H(0, z) = hz$. Next, assume (4.2) and let $u \in f^{t+1}v$. Then

$$u \in f^t y, \qquad y \in fv \qquad \text{for some} \quad y \in N^r$$

Consequently, $u \in (G^{-1}H)y$ and thus

$$
\begin{aligned}
u &= H(x, z'), \qquad y = G(x, z') \\
y &= h(z''), \qquad v = gz''
\end{aligned}
$$

for some z', z'', $x \in N$. Let $z = b(z', z'')$; that is, $z' = cz$, $z'' = dz$. Then

$$u = H(x + 1, z), \qquad v = G(x + 1, z)$$

so that $u \in (G^{-1}H)v$ as required.

To prove the inclusion $G^{-1}H \subset f^+$, we must show that

(4.3) $u = H(x, z)$ and $v = G(x, z)$ imply $u \in f^+v$

This we prove by induction on x. If $x = 0$ then $u = hz$ and $v = gz$ and so $u \in (g^{-1}h)v = fv \subset f^+v$. Next, assume (4.3) for x and let $u = H(x + 1, z)$, $v = G(x + 1, z)$. Then either $u = h0$, $v = g0$ in which case $u \in (g^{-1}h)v = fv \subset f^+v$ or

$$u = H(x, cz), \qquad v = g(dz)$$
$$G(x, cz) = h(dz) = y$$

Since $u = H(x, cz)$, $y = G(x, cz)$, it follows from (4.3) that $u \in f^+y$. Since $y = h(dz)$ and $v = g(dz)$, it follows that $y \in fv$ and thus $u \in f^+v$. This yields (4.3) and together with (4.2) proves (4.1)   ∎

CHAPTER V

Recursive Functions

1. Repetition

We now introduce an operation on partial functions called *repetition*. It assigns to a partial function $h : W^{r+1} \to W^{r+1}$ the partial function $h^\nabla : W^{r+1} \to W^r$ defined as follows. Let $\pi : W^{r+1} \to W$ be the projection given by $\pi(x, w) = w$ for $x \in W^r$, $w \in W$, and let k be the smallest exponent such that $\pi h^k(x, y) = 1$. Then $h^k(x, y) = (x', 1)$ and $h^\nabla(x, y)$ is defined to be x'. If no such exponent exists, then $h^\nabla(x, y) = \varnothing$. We note that since $h^0(x, y) = (x, y)$, we have

$$h^\nabla(x, 1) = x$$

THEOREM 1.1.   *If* $\mathscr{A} \in \mathbf{R}$ *and* $h : W^{r+1} \to W^{r+1}$ *is a partial function in* $\mathscr{A}$, *then* $h^\nabla : W^{r+1} \to W^r$ *is in* $\mathscr{A}$.

Proof.   Define

$$e = \bigcup L_i L_i^{-1} : W \to W, \qquad 1 \leqslant i \leqslant n$$
$$g = (W^r \times e) h : W^{r+1} \to W^{r+1}$$
$$f = W^r \times U^{-1} : W^{r+1} \to W^r$$

Then g and f are in $\mathscr{A}$ and

$$g(x, w) = \begin{cases} h(x, w) & \text{if} \quad w \neq 1 \\ \varnothing & \text{if} \quad w = 1 \end{cases}$$

$$f(x, w) = \begin{cases} \varnothing & \text{if } w \neq 1 \\ x & \text{if } w = 1 \end{cases}$$

Then

$$(g^n f)(x, w) = x'$$

if and only if $h^n(x, w) = (x, 1)$ and n is the lowest exponent for which the second coordinate of $h^n(x, w)$ is 1. Thus $h^\nabla(x, w) = x'$ and consequently, $h^\nabla = g * f$ ∎

Using the operation of repetition, we have the following descriptions of the categories $\mathscr{F}_\Gamma$ and $\mathscr{P}_\Gamma$.

THEOREM 1.2. $\mathscr{F}_\Gamma$ is the smallest admissible subcategory $\mathscr{A}$ of $\mathscr{F}$ containing Γ and such that

(1.1) $\mathscr{A} \in \mathbf{E}$.

(1.2)   If $h : W^{r+1} \to W^{r+1}$ is in $\mathscr{A}$, and $h^\nabla : W^{r+1} \to W^r$ is a function, then h^∇ is in $\mathscr{A}$.

THEOREM 1.2'. $\mathscr{P}_\Gamma$ is the smallest admissible subcategory $\mathscr{A}$ of $\mathscr{P}$ containing Γ and such that

(1.1') $\mathscr{A} \cap \mathscr{F} \in \mathbf{E}$.

(1.2')   If $h : W^{r+1} \to W^{r+1}$ is a function in $\mathscr{A}$, then $h^\nabla : W^{r+1} \to W^r$ is in $\mathscr{A}$.

For the proof, we denote by $\mathscr{F}_\Gamma^\flat$ and $\mathscr{P}_\Gamma^\flat$ the categories defined by the theorems above. Theorem IV,2.1, and Theorem 1.1, respectively, imply that $\mathscr{F}_\Gamma$ satisfies conditions (1.1) and (1.2) while $\mathscr{P}_\Gamma$ satisfies conditions (1.1') and (1.2'). Therefore,

$$\mathscr{F}_\Gamma^\flat \subset \mathscr{F}_\Gamma, \qquad \mathscr{P}_\Gamma^\flat \subset \mathscr{P}_\Gamma$$

The proof of the opposite inclusions is deferred to the next section. We shall remark, however, that it suffices to consider the case $W = N$. Indeed, let $c : W \to N$ be a primitive recursive bijection such that $c1 = 0$. We note that all the categories involved contain $\mathscr{E}_0(W^{\cdot})$ (respectively, $\mathscr{E}_0(N^{\cdot})$) and that the operation of repetition commutes (because of $c1 = 0$) with conjugation. Thus, Proposition III,1.7 may be applied to reduce the argument to the case $W = N$.

2. Minimization

THEOREM 2.1 $(W = N)$. *For any primitive subcategory $\mathscr{A}$ of $\mathscr{F}$, condition (1.2) is equivalent to the following.*

(2.1) *If $f : N^{r+1} \to N$ is a function in $\mathscr{A}$ and $\mu f : N^r \to N$ is a function, then μf is in $\mathscr{A}$.*

THEOREM 2.1' $(W = N)$. *For any admissible subcategory $\mathscr{A}$ of $\mathscr{P}$, such that $\mathscr{A} \cap \mathscr{F} \in \mathbf{E}$, condition (1.2') is equivalent to the following.*

(2.1') *If $f : N^{r+1} \to N$ is a function in $\mathscr{A}$, then $\mu f : N^r \to N$ is in $\mathscr{A}$.*

Proof. We first prove the implications $(1.2) \Rightarrow (2.1)$ and $(1.2') \Rightarrow (2.1')$. Let then a function $f : N^{r+1} \to N$ in $\mathscr{A}$ be given. Define

$$g : N^{r+1} \to N, \qquad g(x, y) = \prod_{t=0}^{y} f(x, t)$$

$$h : N^{r+2} \to N^{r+2}, \qquad h(x, y, z) = (x, 1 + y, g(x, y))$$

for $x \in N^r$, $y, z \in N$. Clearly, $\mu f = \mu g$. Since

$$(2.2) \qquad h^k(x, y, z) = \begin{cases} (x, k + y, g(x, k + y - 1)) & \text{if } k > 0 \\ (x, y, z) & \text{if } k = 0 \end{cases}$$

it follows that

$$h^k(x, 0, g(x, 0)) = (x, k, g(x, k \dot{-} 1))$$

and consequently

$$h^{\nabla}(x, 0, g(x, 0)) = (x, k)$$

where k is the least nonnegative integer such that $g(x, k \mathbin{\dot-} 1) = 0$.
Thus if $\pi : N^{r+1} \to N$ is the projection $\pi(x, y) = y$, we have

$$\pi(h^\nabla(x, 0, g(x, 0))) \mathbin{\dot-} 1 = (\mu g)\, x = (\mu f)\, x$$

We note that h is a function. Further, if μf is a function, then so
is h^∇. Indeed, if $\mu f = \mu g$ is a function, it follows that for every
$x \in N^r$ we have $g(x, y) = 0$ for all y sufficiently large. Thus,
formula (2.2) implies that for fixed (x, y, z), the last coordinate of
$h^k(x, y, z)$ is 0 for all k sufficiently large. Thus, h^∇ is a function.
    Since f is in $\mathscr{A}$ so are g and h. Consequently, h^∇ is in $\mathscr{A}$ and so
is μf.
    Next we prove $(2.1) \Rightarrow (1.2)$ and $(2.1') \Rightarrow (1.2')$. Let then
$h : N^{r+1} \to N^{r+1}$ be a function in $\mathscr{A}$. Define

$$g : N^{r+2} \to N^{r+1}, \qquad g(x, y) = h^y x$$

$$f : N^{r+2} \xrightarrow{g} N^{r+1} \xrightarrow{\pi} N$$

where π is the projection onto the last coordinate. Then $(\mu f)x$ is
the least exponent y for which $\pi(h^y x) = 0$ if such an exponent
exists. Thus

$$g(x, (\mu f)\, x) = (h^\nabla x, 0)$$

Further, if h^∇ is a function, then so is μf. Consequently, if μf is in $\mathscr{A}$,
so is h^∇ ∎

    Given a relation $f : X \to Y$ we call a *selector* for f any partial
function $f' : X \to Y$ such that

$$\operatorname{Dom} f' = \operatorname{Dom} f$$

$$f'x \in fx \qquad \text{for every} \qquad x \in \operatorname{Dom} f$$

PROPOSITION 2.2 ($W = N$). *Every relation* $f : N^r \to N^s$ *in* $\mathscr{R}_\Gamma$
admits a selector of the form

$$N^r \xrightarrow{\mu k} N \xrightarrow{h} N^s$$

with $k : N^{r+1} \to N$, $h : N \to N^s$ *in* $\mathscr{E}_\Gamma$.

Proof. If f is empty, then choose k so that μk is empty (for example, $k(x, y) = 1$) and choose h arbitrarily.

If f is nonempty, then by Corollary IV,3.3, $f = g^{-1}h$ with $g : N \to N^r$, $h : N \to N^s$ in $\mathscr{E}_\Gamma$.

Define $k : N^{r+1} \to N^r$ by setting

$$k(x, y) = |x - gy|$$

Then k is in $\mathscr{E}_\Gamma$. Since $k(x, y) = 0$ holds if and only if $x = gy$ or equivalently if and only if $y \in g^{-1}x$, it follows that

$$(\mu k)\,x = \inf\{g^{-1}x\}$$

where $(\mu k)x = \varnothing$ if the set in braces is empty. Consequently μk is a selector for g^{-1} and thus $(\mu k)h$ is a selector for $g^{-1}h = f$ ∎

We are now ready to complete the proof of Theorems 1.2 and 1.2'; that is,

$$\mathscr{F}_\Gamma \subset \mathscr{F}_\Gamma^\flat, \qquad \mathscr{P}_\Gamma \subset \mathscr{P}_\Gamma^\flat$$

in the case $W = N$. Let then $f : N^r \to N^s$ be in $\mathscr{P}_\Gamma$. Since f is a partial function, it is its only selector and thus in Proposition 2.2 we must have $f = (\mu k)h$. Since μk and h are in $\mathscr{P}_\Gamma^\flat$, it follows that f is in $\mathscr{P}_\Gamma^\flat$. Further, if f is a function, then μk also must be a function. Then μk is in $\mathscr{F}_\Gamma^\flat$ and so is f. *This completes the proof of Theorems 1.2 and 1.2'* ∎

As another consequence of Proposition 2.2, we have

THEOREM 2.3 (selection theorem). *Every relation $f : W^r \to W^s$ in $\mathscr{R}_\Gamma$ admits a selector $f' : W^r \to W^s$ in $\mathscr{P}_\Gamma$.*

Proof. For $W = N$ this follows directly from Proposition 2.2, since μk is in $\mathscr{P}_\Gamma$. The case of a general alphabet follows by conjugation ∎

PROPOSITION 2.4. *Let $f : W^r \to W^s$ be a relation in $\mathscr{R}_\Gamma$ such that $\mathrm{Dom}\, f = W^r$. Then f is the composition*

$$W^r \xrightarrow{\pi^{-1}} W^{r+1} \xrightarrow{g} W^s$$

where g is in $\mathscr{F}_\Gamma$.

Proof. Let $k : W^r \to W^s$ be a selector for f in $\mathscr{P}_\Gamma$. Since Dom $f = W^r$, it follows that Dom $k = W^r$. Thus, k is a function and, therefore, k is in $\mathscr{F}_\Gamma$. From Corollary IV,3.3 we deduce that $f = l^{-1}h$, where $l : W \to W^r$ and $h : W \to W^s$ are in $\mathscr{F}_\Gamma$. Define g by setting

$$g(x, w) = \begin{cases} hw & \text{if} \quad lw = x \\ kx & \text{if} \quad lw \neq x \end{cases}$$

Then by conditional definition g is in $\mathscr{F}_\Gamma$. Further,

$$fx = \{g(x, w) \mid w \in W\}$$

so that $f = \pi^{-1}g$ ∎

3. An Inversion Theorem

THEOREM 3.1. *For every function* $g : W \to W^r$ *in* $\mathscr{F}_\Gamma$, *there exists a partial function* $p : W^{r+1} \to W^{r+1}$ *in* $\mathscr{P}_\Gamma$ *such that the relation* $g^{-1} : W^r \to W$ *is the composition*

$$W^r \xrightarrow{\iota} W^{r+1} \xrightarrow{p^+} W^{r+1} \xrightarrow{\pi} W$$

where ι *is the injection* $\iota x = (x, 1)$ *and* π *is the projection* $\pi(x, y) = y$.

Proof. By conjugation, we need only consider the case $W = N$. Consider the function

$$m : N^{r+2} \to N^{r+2}$$

$$m(x, y, z) = (x, 1 + y, d(x, gy))$$

where $d : N^{r+r} \to N$ is the "distance function"

$$d(y, y') = \sum_{i=1}^{r} |y_i - y_i'|$$

Clearly d is primitive recursive and, therefore, m is in $\mathscr{F}_\Gamma$.

We note that

$$m^{\nabla}(x, y, z) = (x, 1 + \bar{y})$$
$$\bar{y} = \inf\{t \mid gt = x, \quad t \geqslant y\}$$

and $m^{\nabla}(x, y, z) = \varnothing$ if the set in braces is empty.
Define

$$p : N^{r+1} \to N^{r+1}, \qquad p(x, y) = (x, \bar{y})$$

Since m^{∇} is in $\mathscr{P}_r$ and $\bar{y} = \pi(m^{\nabla}(x, y, 0)) \doteq 1$, it follows that p is in $\mathscr{P}_r$. From the definition of $\bar{y}$, it follows that

$$p^+(x, y) = (x, Y)$$
$$Y = \{t \mid gt = x, \quad t \geqslant y\}$$

Therefore,

$$p^+(x, 0) = (x, g^{-1}x)$$

as required ∎

PROPOSITION 3.2. *Given relations*

$$W^r \xrightarrow{p} W^r \xrightarrow{q} W^s$$

the composition p^+q also is the composition

$$W^r \xrightarrow{v} W^{s+r} \xrightarrow{u^+} W^{s+r} \xrightarrow{\pi} W^s$$

with

$$vx = (qx, x), \qquad u(x, y) = (q(py), py)$$

Proof. We have

$$u^n(x, y) = (q(p^n y), p^n y) \qquad \text{for} \quad n > 0$$

If $u^n(x, y) \neq \varnothing$, then $p^n y \neq \varnothing$ and therefore,

$$\pi u^n(x, y) = q(p^n y)) \qquad \text{for} \quad n > 0$$

Consequently, since $vx = (qx, x)$,

$$\pi(u^n(vx)) = q(p^n x)) \qquad \text{for} \quad n > 0$$

It follows that $qp^+ = vu^+\pi$ ∎

PROPOSITION 3.3.   *Every relation $f : W^r \to W^s$ in $\mathscr{R}_\Gamma$ admits a factorization*

$$W^r \xrightarrow{\ l\ } W^{s+r+1} \xrightarrow{\ k^+\ } W^{s+r+1} \xrightarrow{\ \pi\ } W^s$$

with l and k in $\mathscr{P}_\Gamma$ and π a projection.

Proof. If f is empty, we may take k and l to be empty. If not, then $f = g^{-1}h$ with $g : W \to W^r$ and $h : W \to W^s$ in $\mathscr{F}_\Gamma$. Applying Theorem 3.1 to g, we find that f is the composition

$$W^r \xrightarrow{\ \iota\ } W^{r+1} \xrightarrow{\ p^+\ } W^{r+1} \xrightarrow{\ \pi h\ } W^s$$

Thus, applying Proposition 3.2 with $q = p\pi h$, we find that f is the composition

$$W^r \xrightarrow{\ \iota\ } W^{r+1} \xrightarrow{\ v\ } W^{s+r+1} \xrightarrow{\ u^+\ } W^{s+r+1} \xrightarrow{\ \pi\ } W^s$$

Since p and q are in $\mathscr{P}_\Gamma$, so are v and u. This gives the desired result with $l = \iota v$ and $k = u$ ∎

COROLLARY 3.4.   *A relation $f : W^r \to W^s$ is in $\mathscr{R}_\Gamma$ if and only if there exists $t \geqslant s, r$ such that f is the composition*

$$W^r \xrightarrow{\ l\ } W^t \xrightarrow{\ k^+\ } W^t \xrightarrow{\ \pi\ } W^s$$

with l and k in $\mathscr{P}_\Gamma$ ∎

COROLLARY 3.5. $\mathscr{R}_\Gamma$ *is the smallest subcategory of $\mathscr{R}$ containing $\mathscr{P}_\Gamma$ and the closures of partial functions in $\mathscr{P}_\Gamma$ ∎*

4. Distinguished Subcategories of $\mathscr{P}$

So far we have been concerned only with subcategories of $\mathscr{P}$ ($W^\cdot$) that were of the form $\mathscr{P}_\Gamma = \mathscr{R}_\Gamma \cap \mathscr{P}$ where Γ is a class of *functions*. In this section we make a modest attempt to consider other subcategories of $\mathscr{P}$.

PROPOSITION 4.1. *If* $f : W^r \to W^s$, $g : W^r \to W^r$ *are partial functions with disjoint domains, than* $g^+f : W^r \to W^s$ *is a partial function.*

Proof. If suffices to show that $g^n f$ and $g^{n+k} f$ have disjoint domains for $k > 0$. For this, note that

$$\text{Dom } g^{n+k} f \subset \text{Dom } g^{n+1} = (g^n)^{-1} (\text{Dom } g)$$

$$\text{Dom } g^n f = (g^n)^{-1} (\text{Dom } f) \quad \blacksquare$$

A subcategory $\mathscr{A}$ of $\mathscr{P}(W^{\cdot})$ will be called *distinguished* if it is admissible and if it satisfies the following axioms.

(4.1) $U : W^0 \to W$ and $U^{-1} : W \to W^0$ are in $\mathscr{A}$.

(4.2) $L_i : W \to W$ and $L_i^{-1} : W \to W$ are in $\mathscr{A}$ for every $1 \leqslant i \leqslant n$.

(4.3)   If $f, g : W^r \to W^s$ are in $\mathscr{A}$ and have disjoint domains, then $f \cup g : W^r \to W^s$ is in $\mathscr{A}$.

(4.4)   If $f, g : W^r \to W^r$ are in $\mathscr{A}$ and have disjoint domains, then $g^+f : W^r \to W^r$ is in $\mathscr{A}$.

The class of all distinguished subcategories of $\mathscr{P}(W^{\cdot})$ will be denoted by $\mathbf{P}(W^{\cdot})$ or $\mathbf{P}$ for short.

We first prove that axiom (4.4) may be replaced by the slightly stronger axiom.

(4.4′)   If $f : W^r \to W^s$ and $g : W^r \to W^r$ are in $\mathscr{A}$ and have disjoint domains, then so is $g^+f : W^r \to W^s$.

We first consider the case $s < r$. Let then $\iota : W^s \to W^r$, $\pi : W^r \to W^s$ be an injection and a projection such that $\iota\pi = W^s$. Then $g^+f = g^+(f\iota)\pi$ and $g^+(f\iota)$ is in $\mathscr{A}$ by (4.4). In the case $r < s$, we choose an injection $\iota : W^r \to W^s$ and a projection $\pi : W^s \to W^r$ such that $\iota\pi = W^r$. Then $g^+f = \iota(\pi g\iota)^+(\pi f)$ and again (4.4) applies $\blacksquare$

THEOREM 4.2. *In the presence of axioms* (4.1)–(4.3), *axiom* (4.4) *is equivalent to the following.*

(4.4″) *If* $h : W^{r+1} \to W^{r+1}$ *is in* $\mathscr{A}$, *then* $h^\nabla : W^{r+1} \to W^r$ *is in* $\mathscr{A}$.

Proof. For the implication (4.4′) $\Rightarrow$ (4.4″), one needs to examine the proof of Theorem 1.1 and notice that only axioms (4.1), (4.2), (4.3), and (4.4′) were used. To prove the implication (4.4″) $\Rightarrow$ (4.4′), define

$$G : W^{r+1} \to W^{r+1}, \qquad G(x, y) = (gx, \sigma_1)$$

$$F : W^{r+1} \to W^{r+1}, \qquad F(x, y) = (fx, 1)$$

$$h = F \cup G : W^{r+1} \to W^{r+1}$$

Then G, F, and h are in $\mathscr{A}$. Further

$$(g^+f)\, x = h^\nabla(x, \sigma_1)$$

Thus, g^+f is the composition

$$W^r \xrightarrow{W^r \times UL_1} W^{r+1} \xrightarrow{h^\nabla} W^r$$

and, therefore, g^+f is in $\mathscr{A}$ ∎

THEOREM 4.3. *If* $\mathscr{A} \in \mathbf{P}$, *then* $\mathscr{A} \cap \mathscr{F} \in \mathbf{E}$

For this we only need to verify that the proof of Theorem IV,2.1 uses only axioms (4.1), (4.2), (4.3′), and (4.4) ∎

PROPOSITION 4.4. *If* $\mathscr{A} \in \mathbf{R}$, *then* $\mathscr{A} \cap \mathscr{P} \in \mathbf{P}$.

This is clear from the definitions ∎

THEOREM 4.5. $\mathscr{P}_\Gamma$ *is the least element of* $\mathbf{P}$ *containing* Γ.

Proof. Since $\mathscr{P}_\Gamma = \mathscr{R}_\Gamma \cap \mathscr{P}$, it follows from the previous theorem that $\mathscr{P}_\Gamma \in \mathbf{P}$. If $\Gamma \subset \mathscr{A} \in \mathbf{P}$, then by Theorems 4.2 and 4.3, $\mathscr{A}$ satisfies the conditions of Theorem 1.2′. Thus, $\mathscr{P}_\Gamma \subset \mathscr{A}$ ∎

THEOREM 4.6 ($W = N$). $\mathscr{P}_\Gamma$ is the least admissible subcategory $\mathscr{A}$ of $\mathscr{P}$ containing Γ, satisfying axioms (4.1), (4.2), and

(4.4*)   If $f, g : N^r \rightarrow N^r$ are in $\mathscr{A}$, and have disjoint domains, then $g*f : N^r \rightarrow N^r$ is in $\mathscr{A}$.

For the proof, one needs to observe the following facts. The argument establishing the equivalence of (4.4) with (4.4′) remains valid with $+$ replaced by $*$. The proofs of Theorem IV,2.1 and Theorem 1.1 in the case $W = N$ do not use additivity. Thus $\mathscr{A}$ satisfies the conditions of Theorem 1.2′ ∎

CHAPTER VI

Recursively Enumerable Sets

1. $\mathscr{A}$-Enumerable Sets

Given a subset A of W^r we defined $\tau A : W^r \to W^0$ to be the partial function with domain A. It will also be convenient to consider the partial function

$$\rho A : W^r \to W^r, \qquad \rho A = \langle \tau A, W^r \rangle$$

This partial function satisfies

$$(\rho A)\, x = \begin{cases} x & \text{if } x \in A \\ \varnothing & \text{otherwise} \end{cases}$$

Note that τA is the composition

$$W^r \xrightarrow{\ \rho A\ } W^r \xrightarrow{\ \pi\ } W^0$$

PROPOSITION 1.1. *Let $\mathscr{A}$ be a distinguished subcategory of $\mathscr{R}(W^{\cdot})$. Then for any subset A of W^l the following conditions are equivalent:*

(i) *A is the graph of a relation in $\mathscr{A}$,*

(ii) *$\tau A : W^l \to W^0$ is in $\mathscr{A}$,*

(iii) *$\rho A : W^l \to W^l$ is in $\mathscr{A}$,*

(iv) *A is the domain of a partial function in $\mathscr{A}$,*

(v)   *A is the image of a partial function in $\mathscr{A}$,*

(vi) *A is the domain of a relation in $\mathscr{A}$,*

(vii) *A is the image of a relation in $\mathscr{A}$.*

Proof. The equivalence (i) $\Leftrightarrow$ (ii) follows from Proposition IV,1.4. The equivalence (ii) $\Leftrightarrow$ (iii) follows from the formulas relating τA and ρA. The implications (iii) $\Rightarrow$ (iv) $\Rightarrow$ (vi) and (iii) $\Rightarrow$ (v) $\Rightarrow$ (vii) are clear. The equivalence of (vi) and (vii) follows from axiom IV(1.5). Thus it suffices to prove (vi) $\Rightarrow$ (ii). Let A be the domain of $f : W^l \to W^s$. Then τA is the composition $f\pi : W^l \to W^0$ where $\pi : W^s \to W^0$ is the unique function. Since π is in $\mathscr{A}$ it follows that τA is in $\mathscr{A}$ ∎

A subset A of W^l will be called $\mathscr{A}$-*enumerable* if conditions (i)–(vii) are satisfied.

COROLLARY 1.2. *A relation $f : W^r \to W^s$ is in $\mathscr{A}$ if and only if the graph γf is an $\mathscr{A}$-enumerable subset of W^{r+s}.*

Indeed by Proposition IV,1.4 f is in $\mathscr{A}$ if $\tau\gamma f$ is in $\mathscr{A}$ ∎

COROLLARY 1.3. *If $f : W^r \to W^s$ is in $\mathscr{A}$ and $A \subset W^r$ and $B \subset W^s$ are $\mathscr{A}$-enumerable then $fA \subset W^s$ and $f^{-1}B \subset W^r$ are $\mathscr{A}$-enumerable.*

Indeed, $f^{-1}B = \mathrm{Dom}(f(\tau B))$ and $fA = \mathrm{Dom}(f^{-1}(\tau A))$ ∎

We recall that a set $A \subset W^l$ is called an $\mathscr{A}$-set if the characteristic function $\chi A : W^l \to W$ is in $\mathscr{A}$. The characteristic function is defined by $(\chi A)x = \sigma_1$ if $x \in A$ and $(\chi A)x = 1$ if $x \notin A$.

THEOREM 1.4. *A set $A \subset W^l$ is an $\mathscr{A}$-set if and only if both A and its complement $\bar{A}$ are $\mathscr{A}$-enumerable.*

Proof. We note the following

$$\tau A : W^t \xrightarrow{\chi A} W \xrightarrow{L_1^{-1}} W \xrightarrow{U^{-1}} W^0$$
$$\tau \bar{A} : W^t \xrightarrow{\chi A} W \xrightarrow{U^{-1}} W^0$$
$$\chi A = (\tau A)\, U L_1 \cup (\tau \bar{A})\, U$$

Thus, χA is in $\mathscr{A}$ iff both τA and $\tau \bar{A}$ are in $\mathscr{A}$ ∎

THEOREM 1.5 (graph theorem). *For every function* $f : W^r \to W^s$ *the following conditions are equivalent:*

(i)   *f is in $\mathscr{A}$.*

(ii)   *The set $\gamma f \subset W^{r+s}$ is in $\mathscr{A}$.*

(iii)   *The set $\gamma f \subset W^{r+s}$ is $\mathscr{A}$-enumerable.*

Proof. (i) $\Rightarrow$ (ii). This follows from Corollary II,6.5 by conjugation.
(ii) $\Rightarrow$ (iii) follows from Theorem 1.4.
(iii) $\Rightarrow$ (i) follows from Corollary 1.2 ∎

EXERCISE 1.1. *Let* $A \subset W^s$. *Show that the constant relation* $f : W^r \to W^s$ *defined by* $fx = A$ *for all* $x \in W^r$ *is in* $\mathscr{A}$ *if and only if* A *is* $\mathscr{A}$-enumerable.

2. Γ-Enumerable Sets

In the case $\mathscr{A} = \mathscr{R}_\Gamma$ where Γ is a class of functions $W^r \to W^s$, we shall use the phrase Γ-*enumerable* instead of $\mathscr{A}$-enumerable. If $\Gamma = \varnothing$ we shall use "recursively enumerable."

THEOREM 2.1. *For every nonempty subset A of W^t the following properties are equivalent:*

(i)   *A is Γ-enumerable.*

(ii)   *A is the image of a Γ-primitive function $f : W \to W^t$.*

(iii)   *A is the image of a Γ-recursive function $f : W \to W^t$.*

Proof.   (i) $\Rightarrow$ (ii). Since $\tau A : W^l \rightarrow W^0$ is in $\mathscr{R}_\Gamma = \mathscr{E}_\Gamma^{-1}\mathscr{E}_\Gamma$ and τA is not empty we have $\tau A = g^{-1}h$ for some functions $g : W \rightarrow W^l$, $h : W \rightarrow W^0$ in $\mathscr{E}_\Gamma$. Thus $A = \operatorname{Im} g$.
  (ii) $\Rightarrow$ (iii) is obvious.
  (iii) $\Rightarrow$ (i) follows from Proposition 1.1(v)   ∎

  PROPOSITION 2.2.   *Let* $A \subset W^l$ *and let* $\pi : W^{1+l} \rightarrow W^l$ *be the projection. The following conditions are equivalent:*

(i)   *A is Γ-enumerable.*

(ii) $A = \pi B$ *for some Γ-primitive set* $B \subset W^{1+l}$.

(iii) $A = \pi B$ *for some Γ-recursive set* $B \subset W^{1+l}$.

  Proof.   (i) $\Rightarrow$ (ii). If $A = \varnothing$ then $B = \varnothing$ satisfies the conditions. If $A \neq \varnothing$, then by Theorem 2.1, A is the image of a Γ-primitive function $f : W \rightarrow W^l$. Then setting $B = \gamma f$ we have $B \subset W^{1+l}$ and $\pi B = A$. The set B is Γ-primitive by Corollary II,6.5.
  (ii) $\Rightarrow$ (iii) is obvious.
  (iii) $\Rightarrow$ (i). Since B is Γ-recursive, it is also Γ-enumerable. Then $A = \pi B$ is Γ-enumerable by Corollary 1.3   ∎

  PROPOSITION 2.3.   *Let* $A, B \subset W^l$ *be Γ-enumerable sets. There exist then Γ-enumerable sets* $A', B' \subset W^l$ *such that*

$$A' \subset A, \qquad B' \subset B, \qquad A' \cap B' = \varnothing, \qquad A' \cup B' = A \cup B$$

  Proof.   Consider the relation

$$f : W^l \rightarrow W$$

$$f = (\tau A)\, UL_1 \cup (\tau B)\, U$$

which is Γ-recursive. By the selection theorem V,2.3, there exists a Γ-recursive partial function $g : W^l \rightarrow W$ with

$$\operatorname{Dom} g = \operatorname{Dom} f = A \cup B$$

$$gx \in fx \qquad \text{for all} \qquad x \in A \cup B$$

The sets $A' = g^{-1}\sigma_1$, $B' = g^{-1}1$ then have the required properties ∎

3. Axioms for Enumerability

Corollary 1.2 shows that a distinguished subcategory $\mathscr{A}$ of $\mathscr{R}(W^{\cdot})$ is completely determined by the $\mathscr{A}$-enumerable sets. Thus it should be possible to translate the axioms of a distinguished subcategory into the language of sets.

By a *class* of sets $\mathscr{S}$ we shall mean a sequence $\{\mathscr{S}_t\}$, $t \geqslant 0$ where $\mathscr{S}_t$ is a class of subsets of W^t. The class $\mathscr{S}$ will be called *distinguished* if it satisfies the following axioms:

(3.1) If $A, B \in \mathscr{S}_t$ then $A \cap B \in \mathscr{S}_t$.

(3.2) If $A \in \mathscr{S}_t$ then $W \times A \in \mathscr{S}_{t+1}$.

(3.3) If $A \in \mathscr{S}_t$ then $f^\# A \in \mathscr{S}_s$ for every function $f : [s] \to [t]$.

(3.4) $W^0 \in \mathscr{S}_0$.

(3.5) $\{1\} \in \mathscr{S}_1$.

(3.6) $\{(w, \sigma w)\} \in \mathscr{S}_2$ for $\sigma \in \Sigma$.

(3.7) If $A, B \in \mathscr{S}_t$, then $A \cup B \in \mathscr{S}_t$.

(3.8) If $f : W^r \to W^r$ is a relation and $\gamma f \in \mathscr{S}_{2r}$, then $\gamma(f^+) \in \mathscr{S}_{2r}$.

The class of all distinguished classes of sets will be denoted by $\mathbf{S}(W^{\cdot})$.

THEOREM 3.1. *There is a bijection*

$$\mathbf{R}(W^{\cdot}) \approx \mathbf{S}(W^{\cdot})$$

that assigns to each $\mathscr{A} \in \mathbf{R}$ the class of all the graphs of the relations in $\mathscr{A}$ and to each $\mathscr{S} \in \mathbf{S}$ the class of all relations whose graph is in $\mathscr{S}$.

Proof.   Assume $\mathscr{A} \in \mathbf{R}$. The formulas

$$(\rho A)(\rho B) = \rho(A \cap B)$$
$$\rho(W \times A) = W \times \rho A$$
$$[\tau(f^{\#}A)]^{-1} = (\tau A)^{-1}f^{\#}$$
$$\rho W^{0} = W^{0} : W^{0} \to W^{0}$$

imply (3.1)–(3.4). Axioms (3.5)–(3.7) are translations of the axioms IV(1.1)–(1.3). Thus $\mathscr{S} \in \mathbf{S}$.

Conversely let $\mathscr{S} \in \mathbf{S}$ and let $\mathscr{A}$ be the class of all relations whose graph is in $\mathscr{S}$.

We first derive some consequences of (3.1)–(3.4).

(3.9)   If $A \in \mathscr{S}_{t}$, $B \in \mathscr{S}_{s}$, then $A \times B \in \mathscr{S}_{t+s}$.

The set $W^{t} \times B$ is in $\mathscr{S}_{t+s}$ by repeated application of (3.2). The set $A \times W^{s}$ is in $\mathscr{S}_{t+s}$ by an additional application of (3.3), where f is an appropriate permutation. Then

$$A \times B = (A \times W^{s}) \cap (W^{t} \times B)$$

is in $\mathscr{S}_{t+s}$ by (3.1)   ∎

(3.10) $W^{t} \in \mathscr{S}_{t}$ for $t \geqslant 0$

For $t = 0$ this is (3.4). Thus (3.10) follows by a repeated application of (3.2)   ∎

(3.11)   For each $t > 0$ the diagonal set

$$\nabla^{t} = \{(w, \dots, w) \mid x \in W\} \subset W^{t}$$

is in $\mathscr{S}_{t}$.

Indeed, $\nabla^{t} = f^{\#}W$ where $f : [t] \to [1]$ is the unique function   ∎

We are now ready to verify that $\mathscr{A}$ is an admissible subcategory of $\mathscr{R}(W^{\cdot})$ (for this only axioms (3.1)–(3.4) will be used).

Identity. The graph γW^l of the identity function $W^l \to W^l$ is of the form $f^\#(\nabla^2 x \cdots x \nabla^2)$ where $f : [2t] \to [2t]$ is an appropriate permutation. Thus γW^l is in $\mathscr{S}_{2t}$ by (3.11), (3.9), and (3.3).

Composition. Let $f : W^r \to W^s$ and $g : W^s \to W^l$ be in $\mathscr{A}$. Then $\gamma f \in \mathscr{S}_{r+s}$ and $\gamma g \in \mathscr{S}_{s+t}$. Note that

$$\gamma(fg) = h^\#((\gamma f) \times W^t \cap W^r \times (\gamma g))$$

where $h : [r + t] \to [r + s + t]$ is defined by $hi = i$ for $1 \leqslant i \leqslant r$, $hi = s + i$ for $r < i \leqslant r + t$. Thus $\gamma(fg) \in \mathscr{S}_{r+t}$ by (3.9), (3.8), and (3.2). Consequently fg is in $\mathscr{A}$.

Cylindrification. $\gamma(W \times f) = h^\#(\nabla^2 \times \gamma f)$ where h is an appropriate permutation.

Transposition. $\gamma \Theta_k = h^\#(\nabla^2 x \cdots x \nabla^2)$ where $h : [2k] \to [2k]$ is an appropriate permutation.

Diagonal. $\gamma \varDelta = \nabla^3$.

Projection. $\gamma \Pi = W$.

Next we verify axiom IV(1.5) for $\mathscr{A}$. For this we note that $\gamma f^{-1} = h^\#(\gamma f)$ where h is an appropriate permutation.

Axioms IV(1.1)–(1.4) are translations of (3.5)–(3.8). Thus $\mathscr{A} \in \mathbf{R}$.

The arguments above define two functions $\mathbf{R} \to \mathbf{S}$ and $\mathbf{S} \to \mathbf{R}$. The fact that the composition $\mathbf{S} \to \mathbf{R} \to \mathbf{S}$ is the identity is trivial. That the composition $\mathbf{R} \to \mathbf{S} \to \mathbf{R}$ is the identity follows from Corollary 1.2 ∎

COROLLARY 3.2. *If* $\mathscr{A} = \mathbf{R}_\Gamma$, *then* $\mathscr{S}$ *is the smallest admissible class of sets containing* $\{\gamma f\}$ *for* $f \in \Gamma$ ∎

COROLLARY 3.3. *If* $\mathscr{S} \in \mathbf{S}$, *then* $\varnothing \in \mathscr{S}_t$ *for all* $t \geqslant 0$.

Remark. The bijection of Theorem 3.1 remains valid if on the one hand we consider all admissible subcategories of $\mathscr{R}(X^\cdot)$ closed under inversion, and on the other hand all classes of sets satisfying axioms (3.1)–(3.4).

EXERCISE 3.1. *Let* $\varDelta$ *be any class of subsets of* W *and let* $\chi \varDelta$ *denote the class of all characteristic functions* χA *for* $A \in \varDelta$. *Show that the class of* $\chi \varDelta$-*enumerable sets is the smallest element of* $\mathbf{S}(W^\cdot)$ *containing the sets of* $\varDelta$ *and their complements.*

4. Conjugation

Let $W = \Sigma^*$, $V = \Omega^*$ and let

$$c : W \to V$$

be a primitive recursive isomorphism. Let $\mathscr{A} \in \mathbf{R}(W^{\cdot})$ be a distinguished subcategory of $\mathscr{R}(W^{\cdot})$ and let $\mathscr{A}' \in \mathbf{R}(V^{\cdot})$ be the conjugate distinguished subcategory of $\mathscr{R}(\mathbf{V})$. Let $\mathscr{S} \in \mathbf{S}(W^{\cdot})$ be the class of $\mathscr{A}$-enumerable sets and $\mathscr{S}' \in \mathbf{S}(V^{\cdot})$ the class of $\mathscr{A}'$-enumerable sets.

For any relation $f : W^r \to W^s$ we note the identity

$$\gamma(f') = c(\gamma f)$$

Furthermore, for any subset A of W^r we have

$$\tau(cA) = (\tau A)'$$

This yields

PROPOSITION 4.1. $\mathscr{S}' = c\mathscr{S}$ ∎

This symbolic equation says that $A' \in \mathscr{S}'_t$ iff $A' = cA$ for some $A \in \mathscr{S}_t$.

COROLLARY 4.2.   *A recursive bijection* $c : W \to V$ *establishes a bijection*

$$\mathbf{S}(W^{\cdot}) \approx \mathbf{S}(V^{\cdot})$$

and this bijection is independent of the choice of c.

This follows from Theorem IV,2.3   ∎

A very important special case obtains when Σ is a subset of Ω. Then W is a submonoid of V; let

$$i : W \to V$$

be the inclusion morphism.

PROPOSITION 4.3. *The functions*

$$W \xrightarrow{\ i\ } V \xrightarrow{\ c^{-1}\ } W$$
$$V \xrightarrow{\ c^{-1}\ } W \xrightarrow{\ i\ } V$$

are primitive recursive.

Proof. Since $i(\sigma w) = \sigma(iw)$, we have $L_\sigma i = iL_\sigma$. Setting $g = ic^{-1}$ we obtain

$$L_\sigma g = L_\sigma ic^{-1} = iL_\sigma c^{-1} = ic^{-1}cL_\sigma c^{-1} = gL_\sigma'$$

Equivalently

$$g(\sigma w) = L_\sigma'(gw)$$

Thus by induction, g is primitive recursive. Since

$$c^{-1}i = c^{-1}ic^{-1}c = c^{-1}gc = g'$$

the primitive recursiveness of $c^{-1}i$ follows ∎

PROPOSITION 4.4. *If $B \subset V^l$ is in $\mathscr{S}'$, then $B \cap W^l \subset W^l$ is in $\mathscr{S}$. A set $A \subset W^l$ is in $\mathscr{S}$ if and only if as a subset of V^l it is in S'.*

Proof. Let $g = ic^{-1}$, $h = c^{-1}i$. Then

$$B \cap W^t = i^{-1}B = (i^{-1}cc^{-1})\,B = (h^{-1}c^{-1})\,B = c^{-1}(h^{-1}B)$$

If B is in $\mathscr{S}'$, then so is $h^{-1}B$ since h is primitive recursive. Thus $B \cap W^l$ is in $\mathscr{S}$.
    Further,

$$iA = (ic^{-1}c)\,A = (gc)\,A = c(gA)$$

If A is in $\mathscr{S}$, then so is gA, since g is primitive recursive. Then $c(gA)$ is in $\mathscr{S}'$ and $iA = A$ is in $\mathscr{S}'$ ∎

COROLLARY 4.5. *A relation $f: W^r \to W^s$ is in $\mathscr{A}$ if and only if the relation*

$$V^r \xrightarrow{\ i^{-1}\ } W^r \xrightarrow{\ f\ } W^s \xrightarrow{\ i\ } V^t$$

is in $\mathscr{A}'$.

This follows from the identity

$$\gamma(i^{-1}fi) = i(\gamma f) \quad \blacksquare$$

COROLLARY 4.6.   *A relation* $f : W^r \to W^s$ *is in* $\mathscr{A}$ *if and only if f is the composition*

$$W^r \xrightarrow{\ i\ } V^r \xrightarrow{\ g\ } V^s \xrightarrow{\ i^{-1}\ } W^s$$

for some g in $\mathscr{A}'$.

This follows from the identities

$$f = i(i^{-1}fi)\,i^{-1}$$
$$\gamma(igi^{-1}) = \gamma(g) \cap W^{r+s} \quad \blacksquare$$

EXERCISE 4.1.   *Let* $f : W \to V$ *be a morphism of monoids. Prove the analog of Proposition 4.3 with i replaced by f. Show that if* $A \subset W^r$ *is recursively enumerable then so is* $fA \subset V^r$. *Show that if* $B \subset V^r$ *is recursively enumerable then so is* $f^{-1}B \subset W^r$. *Formulate analogous statements for recursive relations.*

A Recursive Function
Which Is Not Primitive Recursive

The purpose of this appendix is to show that $\mathscr{E}_0 \neq \mathscr{F}_0$.

Before proceeding with the actual example, we shall give an argument that will illuminate some aspects of the proof. Assume that $\mathscr{E}_0 \neq \mathscr{F}_0$. It then follows from Theorem V,2.1 that the set of primitive recursive functions is not closed under minimization. Thus, there exists a primitive recursive function $f : N^{r+1} \to N$ such that $\mu f : N^r \to N$ is a function and is not primitive recursive. Further, thanks to the bijection $N^r \approx N$, we may assume that $r = 1$. The function $\mu f : N \to N$ is then recursive but is not dominated by any primitive recursive function, since otherwise, μf itself would be primitive recursive by dominated minimization. Thus the recursive function μf "grows faster" than any primitive recursive function. This tends to explain the fact that the proof that follows will use a "rate of growth" argument.

In II,8 we asserted that the function

(A.1) $$\Psi : N^2 \to N$$

satisfying

(A.2)
$$\Psi(0, y) = y + 1$$
$$\Psi(x + 1, 0) = \Psi(x, 1)$$
$$\Psi(x + 1, y + 1) = \Psi(x, \Psi(x + 1, y))$$

is recursive without being primitive recursive. First we must

73

verify that the above conditions indeed define a function (A.1). To see this we write $\Psi_n(y)$ instead of $\Psi(n, y)$. Then the third condition becomes

$$\Psi_{n+1}(y + 1) = \Psi_n(\Psi_{n+1}(y))$$

Iteration yields

$$\Psi_{n+1}(y + 1) = \Psi_n^{y+1}(\Psi_{n+1}(0))$$

Since by the second condition $\Psi_{n+1}(0) = \Psi_n(1)$, we obtain

$$\Psi_{n+1}(y + 1) = \Psi_n^{y+2}(1)$$

Thus the sequence of function

$$\Psi_n : N \to N, \qquad n \geqslant 0$$

may be defined inductively by the formulas

(A.3)
$$\Psi_0(y) = y + 1$$
$$\Psi_{n+1}(y) = \Psi_n^{y+1}(1)$$

The second of these conditions may also be written as

$$\Psi_{n+1}(y) = \Psi_n^{\S}(1, y + 1)$$

which shows that the functions Ψ_n are primitive recursive for $n \geqslant 0$.

We leave it to the reader to verify (by induction) the inequalities

(A.4) $$\Psi(x, y) < \Psi(x, y + 1) \leqslant \Psi(x + 1, y)$$

which in particular show that the function Ψ is strictly ascending in both variables.

Given $x = (x_1, ..., x_r) \in N^r$ we define

$$|x| = x_1 + \cdots + x_r$$

In particular, $|0| = 0$. If $r = 0$, then the unique element $0 \in N^0$ has norm zero.

PROPOSITION A.1. *For every primitive recursive function* $f : N^r \to N^s$ *there exists an integer* $m \geq 0$ *such that*

(A.5) $| fx | < \Psi(m, | x |)$

for all $x \in N^r$.

Proof. Let $\mathcal{B}_m$ denote the class of all functions $f : N^r \to N^s$ (for all $r \geq 0$, $s \geq 0$) for which (A.5) holds. In view of (A.4) we have $\mathcal{B}_m \subset \mathcal{B}_{m+1}$. Let $\mathcal{B}$ be the union of the classes $\{\mathcal{B}_m\}$, $m \geq 0$. To prove Proposition A.1 it suffices to show that $\mathcal{B}$ is a primitive subcategory of $\mathcal{F}(N^{\cdot})$.

If $f : N^r \to N^r$ is the identity, then $| fx | = | x | < \Psi(0, | x |)$, so that $f \in \mathcal{B}_0$.

Consider functions

$$N^r \xrightarrow{f} N^s \xrightarrow{g} N^l$$

in $\mathcal{B}_m$. Then

$$
\begin{aligned}
|(fg)\, x | = | g(fx)| &\leq \Psi(m, | fx |) \\
&\leq \Psi(m, \Psi(m, | x |)) \\
&< \Psi(m, \Psi(m + 1, | x |)) \\
&= \Psi(m + 1, | x | + 1) \\
&\leq \Psi(m + 2, | x |)
\end{aligned}
$$

so that $fg \in \mathcal{B}_{m+2}$. This shows that $\mathcal{B}$ is a subcategory of $\mathcal{F}(N^{\cdot})$.

Next we consider the basic logical functions. The transposition $\Theta_k : N^k \to N^k$ satisfies $| \Theta_k x | = | x |$. Thus $| \Theta_k x | \leq \Psi(0, | x |)$ and therefore, Θ_k is in $\mathcal{B}_0$. The projection $\Pi : N \to N^0$ satisfies $| \Pi x | \leq \Psi(0, | x |)$ and is in $\mathcal{B}_0$. The diagonal $\varDelta : N \to N^2$ satisfies $| \varDelta x | = |(x, x)| = 2x$. We let the reader verify the formula $\Psi(2, x) = 3 + 2x$ which implies that $\varDelta$ is $\mathcal{B}_2$.

Let $f \in \mathcal{B}_m$ and let $g = Nxf$ be the cylindrification of f. Then

$$
\begin{aligned}
| g(n, x)| = |(n, fx)| = n + | fx | & \\
&\leq n + \Psi(m, | x |) \\
&\leq \Psi(m, n + | x |) \\
&= \Psi(m, |(n, x)|)
\end{aligned}
$$

so that $g \in \mathscr{B}_m$. We have shown thus far that $\mathscr{B}$ is an admissible subcategory of $\mathscr{F}(N^*)$.

Since $S = \Psi_0$ we have $S \in \mathscr{B}_0$. Similarly $U \in \mathscr{B}_0$. There remains to be verified that $\mathscr{B}$ is closed under exponentiation. Let then $f : N^r \to N^r$ be in $\mathscr{B}_m$. We assert that $f^s \in \mathscr{B}_{m+1}$, that is, that

(A.6) $|f^i x| \leqslant \Psi(m + 1, |x| + i)$

For $i = 0$ we have

$$|f^0 x| = |x| \leqslant \Psi(m + 1, |x|)$$

so that (A.6) holds. Now assume (A.6) and consider $f^{i+1} x$. We have

$$
\begin{aligned}
|f^{i+1} x| = |f(f^i x)| &\leqslant \Psi(m, |f^i x|) \\
&\leqslant \Psi(m, \Psi(m + 1, |x| + i)) \\
&= \Psi(m + 1, |x| + i + 1) \quad \blacksquare
\end{aligned}
$$

COROLLARY A.2.   *The function*

$$f : N \to N, \qquad fx = \Psi(x, x)$$

is not primitive recursive.

Indeed, if (A.5) holds for f, then

$$\Psi(m + 1, m + 1) = f(m + 1) \leqslant \Psi(m, m + 1)$$

contradicting (A.4) $\blacksquare$

COROLLARY A.3.   *The function Ψ is not primitive recursive* $\blacksquare$

We now proceed with the proof that Ψ is recursive. Let $X \subset N^3$ be the graph of Ψ. By Theorem VI,1.5 it suffices to show that X is recursively enumerable. We note the following properties of X

$$(0, x, x + 1) \in X$$

(A.7) $(x, 1, y) \in X \Rightarrow (x + 1, 0, y) \in X$

$$(x + 1, y, z) \in X \& (x, z, t) \in X \Rightarrow (x + 1, y + 1, t) \in X$$

The above conditions are a translation of conditions (A.2) and it is clear that X is the least subset of N^3 satisfying (A.7).
    Define

$$A = \{(0, x, x + 1)\} \subset N^3$$

$$f_1 : N^3 \to N^3$$

$$f_1(x, 1, y) = (x + 1, 0, y)$$

$$f_2 : N^6 \to N^3$$

$$f_2((x + 1, y, z), (x, z, t)) = (x + 1, y + 1, t)$$

with the understanding that f_1 and f_2 are undefined except when specified above. Clearly f_1 and f_2 are recursive partial functions, while the set A is recursively enumerable. Further the set X is the least solution of the inclusion

(A.8) $$A \cup f_1 X \cup f_2(X, X) \subset X$$

The specific nature of A, f_1, and f_2 is no longer of interest. Further, an application of the recursive isomorphism $N^3 \approx N$ permits the replacement of N^3 by N.
    If we introduce the relation $f_0 : N^0 \to N$ which to the unique element of N^0 assigns the set A, then f_0 is a recursive relation and (A.8) may be rewritten as

$$f_0 X^0 \cup f_1 X^1 \cup f_2 X^2 \subset X$$

where X^i stands for the i-fold product $X \times \cdots \times X$ and $X^0 = N^0$. The required result now follows from

PROPOSITION A.4. *Let*

$$f_i : N^i \to N, \qquad 0 \leqslant i \leqslant u$$

be recursive relations. Then the least solution of the inclusion

(A.9) $$\bigcup_{i=0}^{u} f_i X^i \subset X$$

is a recursively enumerable subset of N.

Proof. Let N^ω denote the set of all finite sequences

$$x = (x_1, ..., x_p), \qquad p \geqslant 0$$

of elements of N. Thus

$$N^\omega = N^0 \cup N^1 \cup \cdots \cup N^p \cup \cdots$$

and in this context the unique element of N^0 will be written as the 0-tuple ( ). We consider the partial functions

$$R : N^\omega \to N^\omega, \qquad R(x_1, ..., x_p) = (x_1, ..., x_{p-1})$$
$$T : N^\omega \to N, \qquad T(x_1, ..., x_p) = x_p$$

with R and T undefined on ( ). Thus if $x \in N^\omega$ and $x \neq$ ( ) then

$$x = (Rx, Tx)$$

Given $x \in N^\omega$ and $n \in N$ we shall write

$$x \Rightarrow n$$

if $x = (x_1, ..., x_p)$, $p \geqslant 0$ and one of the following two conditions holds:

(i) $n = x_i$ for some $1 \leqslant i \leqslant p$,

(ii) $n \in f_j(y_1, ..., y_j)$ for some $0 \leqslant j \leqslant t$ and some $y_1, ..., y_j \in \{x_1, ..., x_p\}$.

The sequence $(x_1, ..., x_p)$ is called a *derivation* of length p if

$$(x_1, ..., x_{i-1}) \Rightarrow x_i$$

for all $1 \leqslant i \leqslant p$. In particular if $p > 0$, it follows that $x_1 \in f_0($). Note that the sequence ( ) is a derivation. Let D_p denote the set of all derivations of length p, and let $D = D_0 \cup D_1 \cup \cdots \cup D_p \cup \cdots$. We assert that

(A.10) $X = TD$

For the proof we note

(iii) $TD_p \subset TD_{p+1}$.

(iv) $f_i(TD_p ,..., TD_p) \subset D_{ip+1}$.

To prove (iii), it suffices to observe that if $(x_1 ,..., x_p)$ is a derivation, then so is $(x_1 ,..., x_p , x_p)$. To prove (iv), assume that

$$(x_{j1} ,..., x_{jp}),   1 \leqslant j \leqslant i$$

are derivations and that $y \in f_i(x_{1p} ,..., x_{ip})$. Then

$$(x_{11} ,..., x_{1p} ,..., x_{i1} ,..., x_{ip} , y)$$

is a derivation of length $ip + 1$. This proves (iv). From (iii) and (iv) it follows that TD satisfies the inclusion (A.9) and that $X \subset TD$. To prove the opposite inclusion we prove by induction on p that $TD_p \subset X$. For $p = 0$ this is clear since $TD_0 = \varnothing$. Assume $TD_p \subset X$ and let $(x_1 ,..., x_p , x_{p+1})$ be a derivation. Then $(x_1 ,..., x_p) \in D_p$ and thus $x_1 ,..., x_p \in X$ by the inductive hypothesis. Further, $(x_1 ,..., x_p) \Rightarrow x_{p+1}$. Now from (i) and (ii) it follows that $x_{p+1} \in X$. Thus $TD_{p+1} \subset X$.

Next we introduce the relations

$$G : N^\omega \to N^\omega,   H : N^\omega \to N$$

by setting

$$y \in Gx   \text{iff}   y = (x, n) \,\&\, x \Rightarrow n$$

$$n \in Hx   \text{iff}   x \Rightarrow n$$

The following formulas are then clear:

(A.11)
$$TD = G^*T(\ )$$
$$G = R^{-1} \cap HT^{-1}$$

Finally we introduce the relations

$$K, L_i : N^\omega \to N,   1 \leqslant i \leqslant u$$

by setting

$$Kx = \{x_1, ..., x_p\} \qquad \text{if} \quad x = (x_1, ..., x_p)$$

and defining L_i as the composition

$$N^\omega \xrightarrow{\quad \Delta \quad} N^{\omega i} \xrightarrow{\quad K^i \quad} N^i \xrightarrow{\quad f_i \quad} N$$

where Δ is an appropriate diagonal map. With these definitions
we note the formulas

(A.12)
$$K = R*T$$
$$H = K \cup L_0 \cup \cdots \cup L_u$$

The next element needed in the proof is a "coding," that is,
a bijection

$$c : N^\omega \twoheadrightarrow N$$

satisfying the following three conditions

$$c(\ ) = 0$$

(A.13)
$$r : c^{-1}Rc : N \to N \quad \text{is recursive}$$
$$t = c^{-1}T : N \to N \quad \text{is recursive}$$

Assuming that such a coding is obtained, we set

$$g = c^{-1}Gc, \qquad h = c^{-1}H$$
$$k = c^{-1}K, \qquad l_i = c^{-1}L_i$$

Then since $K = R*T$, we have $k = r*t$ and therefore k is recursive.
Since l_i is the composition

$$N \xrightarrow{\quad \Delta \quad} N^i \xrightarrow{\quad k^i \quad} N^i \xrightarrow{\quad f_i \quad} N$$

it follows that $l_0, ..., l_u$ are recursive. Formula

$$h = k \cup l_0 \cup \cdots \cup l_u$$

then implies that h is recursive. Since $g = r^{-1} \cap ht^{-1}$ it follows that g is recursive. Finally since $TD = (g*t)0$ it follows that TD is recursively enumerable.

To conclude the proof we must exhibit a coding $c : N^\omega \to N$ satisfying (A.13). For this purpose we set

$$c(\ ) = 0$$

$$cx = b(c(Rx), Tx) + 1 \quad \text{if} \quad x \neq (\ )$$

where $b : N^2 \to N$ is the bijection of II,7. If $cx = cy$, then it follows that $Tx = Ty$ and $c(Rx) = c(Ry)$. Thus it follows by induction that $x = y$. Consequently c is injective. To prove that c is surjective let $n \in N$. If $n = 0$, then $c(\ ) = n$. If $n > 0$, then $n = b(m, k) + 1$ for some $m, k \in N$. From II(7.4) we deduce that $m < n$ and thus by induction we may assume that $m = cy$ for some $y \in N^\omega$. But then

$$n = b(cy, k) + 1 = c(y, k)$$

With c thus defined the partial functions $c^{-1}Rc$ and $c^{-1}T$ become the compositions

$$N \xrightarrow{S^{-1}} N \xrightarrow{b^{-1}} N \times N \xrightarrow{\pi_i} N, \quad i = 1, 2$$

where π_i are the projections ∎

The function Ψ was introduced by Péter (Reference 5, p. 106).

Degrees

The methods of this monograph lead to a somewhat novel approach to the problems connected with degrees of unsolvability. Without getting involved in the details of this quite elaborate subject, we shall be satisfied here with a formulation of some of the basic definitions.

The set **D** of *degrees* is defined as the subset of $\mathbf{R}(W^{\cdot})$ given by the distinguished subcategories of $\mathbf{R}(W^{\cdot})$ which are of the form $\mathscr{R}_\Gamma$ where Γ is a *finite set of functions*.

In view of the bijection $\mathbf{R}(W^{\cdot}) \approx \mathbf{R}(N^{\cdot})$ we may assume $W = N$ throughout. Since the equalities

$$\mathscr{F}_\Gamma = \mathscr{F}_{\Gamma'}, \qquad \mathscr{R}_\Gamma = \mathscr{R}_{\Gamma'}$$

are equivalent (by the main theorem), it follows that **D** may be viewed either as a subset of **F** or as a subset **R** whichever is more convenient.

Next we observe that if $\Gamma = \{f_1, ..., f_k\}$ and $f = f_1 \times \cdots \times f_k$ then $\mathscr{R}_\Gamma = \mathscr{R}_f = \mathscr{R}_{x \gamma f}$. Further, using the bijections $N^r \approx N$, one may replace γf by a subset A of N. Thus each degree has the form $\mathscr{R}_{xA}$, where A is a subset of N. If we pass to the corresponding distinguished class of sets, we obtain the least distinguished class of sets containing A and $\bar{A}$. Those degrees for which A may be chosen to be recursively enumerable yield an important subset **C** of **D**.

The set $\mathbf{D}$ is ordered by inclusion and its smallest element is the category $\mathscr{R}_0$ of recursive relations. It is further clear that $\mathbf{D}$ is an upper semilattice. Indeed, $\mathscr{R}_{\Gamma_1 \cup \Gamma_2}$ is the least degree containing both $\mathscr{R}_{\Gamma_1}$ and $\mathscr{R}_{\Gamma_2}$.

We conclude with a few simple facts which follow from the remark that $\mathscr{R}_\Gamma$ is countable if Γ is finite (or countable). If the degree d is $\mathscr{R}_\Gamma$ then every $d' < d$ is given by a finite subset Γ' of $\mathscr{R}_\Gamma$. Since $\mathscr{R}_\Gamma$ is countable we obtain the fact that the set

$$\{d' \mid d' < d\}$$

is at most countable for every degree d. In contrast however the set $\mathbf{D}$ itself is uncountable. In fact, $\mathbf{D}$ contains no countable subset that is cofinal. Indeed, let $\{d_i\}$ be a sequence of degrees that is cofinal, and let $d_i = \mathscr{R}_{\Gamma_i}$. Then $\Gamma = \bigcup_i \Gamma_i$ is countable and so is $\mathscr{R}_\Gamma$. The sequence $\{d_i\}$ being cofinal implies that $\mathscr{R}_\Gamma = \mathscr{R}$, a contradiction since $\mathscr{R}$ is uncountable.

Since the class of all recursively enumerable sets is countable, the subset $\mathbf{C}$ of $\mathbf{D}$ is countable.

For further results about degrees, see Rogers (Reference 6, Chapters 9–13).

Bibliography

1. Davis, M., "Computability and Undecidability." McGraw–Hill, New York, 1958.
2. Davis, M., ed., "The Undecidable." Raven Press, Hewlitt, New York, 1965.
3. Hermes, H., "Enumerability, Decidability, Computability." Springer, Berlin, 1965.
4. Kleene, S., "Introduction to Metamathematics." Van Nostrand, Princeton, New Jersey, 1952.
5. Péter, R., "Recursive Functions," 3rd edition. Academic Press, New York, 1967.
6. Rogers, H., "Theory of Recursive Functions and Effective Computability." McGraw–Hill, New York, 1967.

Index

Action
  left, 11
  right, 12
Admissible subcategory, 5

Category, 3
  composition in, 3
  large, 4
  morphisms of, 3
  objects of, 3
  small, 5
Characteristic function, 20
Class of sets, 67
  distinguished, 67
Closure, 39
Conditional definition, 18
Conjugate, 29
Cylindrification, 5

Degree, 83
Diagonal, 6
Domain, 3

Enumerable sets, 64
  recursively, 65
Exponential
  left, 12
  right, 13

Free monoid, 8
  letters of, 8
  words of, 8

Graph, 2

Induction, 15

Logical function, 6
  basic, 6

Minimization, 18
  dominated, 19
  of set, 21
Morphisms, 3

Objects of category, 3

Partial function, 3
Primitive recursive
  function, 14
  isomorphism, 23, 30
  relative to Γ, 15
  sets, 20
Primitive subcategory, 13
Projection, 6

Recursively enumerable, 65
Recursive relation, 44
Relation, 2
  domain of, 3
  graph of, 2
Repetition, 51
Reversal, 8

Selector, 54
Subcategory, 4
  admissible, 5
  distinguished, 39
  full, 4
  primitive, 13
Successor, 8
  left, 8
  right, 8

Transposition, 6

Unit, 8

Words, 8
  length of, 8

Index of Notation

$\hat{X}$	2		ρ	8
$\varnothing$	2		S	9
$\mathrm{Dom}\, f$	2		$\mid w \mid$	9
$\mathrm{Obj}\, \mathscr{A}$	3		$^{\S}(k_1 ,..., k_n)$	$\cdot$12
$\mathscr{A}(A, B)$	3		$(k_1 ,..., k_n)^{\S}$	13
1_A	4		$\mathbf{E}(W^{\cdot})$	14
$\mathscr{R}$	4		$\mathscr{E}_0(W^{\cdot})$	14
$\mathscr{P}$	4		$\mathscr{E}_{\Gamma}(W^{\cdot})$	15
$\mathscr{F}$	4		f^{+}	39
X^r	5		$\mathbf{R}(W^{\cdot})$	40
$X^{\cdot}$	5		τA	40
$\mathscr{R}(X^{\cdot})$	5		f^{*}	41
$\mathscr{P}(X^{\cdot})$	5		$\mathscr{R}_{\Gamma}$	44
$\mathscr{F}(X^{\cdot})$	5		$\mathscr{P}_{\Gamma}$	44
(A, B)	5		$\mathscr{F}_{\Gamma}$	44
$X \times f$	5		$\mathscr{R}_0$	44
Θ	6		$\mathscr{P}_0$	44
Δ^k	6		$\mathscr{F}_0$	44
Π	6		$\mathscr{A}^{-1}\mathscr{A}$	44
$[n]$	6		$\mathrm{Im}\, k$	45
$f^{\#}$	6		$\mathbf{P}$	59
$f \times g$	7		ρA	63
$\langle f, g \rangle$	7		$\mathbf{S}(W^{\cdot})$	67
L_i	8		$\mathbf{D}$	83
R_i	8		$\mathbf{C}$	83
U	8			